Evolutionary Neuropsychology

Evolutionary Neuropsychology

An Introduction to the Evolution of the Structures and Functions of the Human Brain

FREDERICK L. COOLIDGE, PHD

OXFORD
UNIVERSITY PRESS

OXFORD
UNIVERSITY PRESS

Oxford University Press is a department of the University of Oxford. It furthers
the University's objective of excellence in research, scholarship, and education
by publishing worldwide. Oxford is a registered trade mark of Oxford University
Press in the UK and certain other countries.

Published in the United States of America by Oxford University Press
198 Madison Avenue, New York, NY 10016, United States of America.

Library of Congress Cataloging-in-Publication Data
Names: Coolidge, Frederick L. (Frederick Lawrence), 1948– author.
Title: Evolutionary neuropsychology : an introduction to the evolution of
the structures and functions of the human brain / Frederick L. Coolidge, PhD,
Psychology Department, University of Colorado, Colorado Springs.
Description: New York : Oxford University Press, [2020] |
Includes bibliographical references and index.
Identifiers: LCCN 2019033312 (print) | LCCN 2019033313 (ebook) |
ISBN 9780190940942 (hardback) | ISBN 9780190940966 (epub) |
ISBN 9780190940959 (updf) | ISBN 9780190940973 (online)
Subjects: LCSH: Neuropsychology. | Brain—Evolution.
Classification: LCC QP360.C6674 2020 (print) | LCC QP360 (ebook) |
DDC 612.8/233—dc23
LC record available at https://lccn.loc.gov/2019033312
LC ebook record available at https://lccn.loc.gov/2019033313

1 3 5 7 9 8 6 4 2

Printed by Integrated Books International, United States of America

Contents

Preface

Evolutionary neuropsychology is a scientific specialty focusing on the ultimate, evolutionary origins of the brain's structures and their specialized functions. This book is designed for the intellectually curious, but styled especially for academics at any level and psychologists focusing on various aspects of human behavior. It is the first of its kind in a new multidisciplinary science that embraces and uses empirical findings from the fields of evolution, neuroscience, cognitive neuroscience, psychology, anthropology, and archaeology. The bedrock foundation of evolutionary neuropsychology is the assumption that functionally specialized brain regions are adaptations naturally selected in response to various environmental challenges over the course of billions of years of evolution. These adaptations and their brain regions and circuitry may now serve new functions, which are called *exaptations,* and they are particularly involved in higher cognitive functions. As Charles Darwin noted in 1862, "Although an organ may have not been originally formed for some special purpose, if it now serves for this end, we are justified in saying that it is specially adapted for it. . . . Thus throughout nature almost every part of each living being has probably served, in a slightly modified condition, for diverse purposes, and has acted in the living machinery of many ancient and distinct specific forms" (p. 348). Thus, Darwin was the first to recognize that exaptations of organs may come to serve a different purpose. A classic example of an exaptation is feathers. Feathers were originally selected for the purposes of thermoregulation. Millions of years of genetic mutations later, some creatures co-opted feathers for flight and the faster mobility that they afforded. This exaptation gave these creatures a reproductive fitness advantage over those that were slower and flightless. Employing evolutionary neuropsychology's foundational position, this book presents evidence that the functions of genetically or epigenetically derived neurons and neural circuits were either repurposed or reused to serve new and higher cognitive functions that uniquely characterize modern human brains.

The book begins with an overview of the origins of life and brains, and presents a brief history of life's timeline from the coagulation of molecules, to the first unicellular organisms, to modern *Homo sapiens*. It continues with a provocative chapter (Chapter 2) on the origins of learning and memory systems, tracing those essential systems back to their ultimate origin: chemical affinities and disaffinities, which are themselves based on properties and activities of subatomic particles. The book then pursues these neural adaptations and exaptations through the major lobes of the brain—frontal, parietal, temporal, and occipital—and through other regions (i.e., cerebellum and hippocampus) critical to higher cognitive functions like language, reasoning, decision-making, and theory of mind (i.e., appropriately understanding the attitudes, intentions, and feelings of others). The penultimate chapter (Chapter 9) covers the last great leap of humans and their brains, which began about 2 million years ago when a new species transitioned from arboreal to full terrestrial life. This dramatically new species was *Homo erectus*, who began the transition from an ape-like way of living to a more human-like way of life. The final chapter (Chapter 10), on paleopsychopathogy, discusses common, disabling, and maladaptive mental disorders, such as ADHD, PTSD, personality disorders, and schizophrenia, from the perspective that they may have had adaptive value in the ancestral environment but are maladaptive in the present environment.

List of Abbreviations

A	adenine
AASM	American Academy of Sleep Medicine
ACC	anterior cingulate cortex
ADD	attention-deficit disorder (archaic)
ADHD	attention-deficit hyperactivity disorder
AG	angular gyrus
AIC	anterior insular cortex
APCs	action perception circuits
ASD	autism spectrum disorder
aSMG	anterior supramarginal gyrus
BA	Brodmann's area
C	cytosine
CCAS	cerebellar cognitive affective syndrome
CIA	Central Intelligence Agency
CR	conditioned response
CS	conditioned stimulus
DLPFC	dorsolateral prefrontal cortex
DMN	default mode network
DNA	deoxyribonucleic acid
DSM	*Diagnostic and Statistical Manual of Mental Disorders*
DSM-5	*Diagnostic and Statistical Manual of Mental Disorders* (5th ed.)
E. coli	*Escherichia coli*
EEA	environment of evolutionary adaptiveness
EEG	electroencephalogram
EQ	encephalization quotient
ESS	evolutionary stable strategies
FBI	Federal Bureau of Investigation
FFA	fusiform face area
fMRI	functional magnetic resonance imaging
G	guanine
hIPS	horizontal intraparietal sulcus
H.M.	patient H.M.
IPL	inferior parietal lobule
IPS	intraparietal sulcus
IQ	intelligence quotient
K-selection	Kapazitätsgrenze (capacity limit)-selection.

MBD	minimal brain dysfunction (archaic)
MCPH1	microcephalin gene
MDR	massive redeployment hypothesis
MPI	male parental investment
MS	multiple sclerosis
NOTCH	a group of genes
NOTCH 2NL	three specific genes in the NOTCH group
N1	stage 1 sleep
N2	stage 2 sleep
N3	stage 3 and 4 sleep
O	object (in a sentence)
OFC	orbitofrontal cortex
OFPFC	orbitofrontal prefrontal cortex
OSV	object subject verb (word order)
OVS	object verb subject (word order)
PAS	precategorical acoustic storage
PET	positron emission tomography
PFC	prefrontal cortex
POT	parietal occipital temporal (junction)
PTSD	posttraumatic stress disorder
R	REM sleep
REM	rapid eye movement
RNA	ribonucleic acid
RSC	retrosplenial cortex
r-selection	rate-selection
rSMG	right supramarginal gyrus
S	subject (in a sentence)
SMG	supramarginal gyrus
SMGp	supramarginal gyrus posterior
SOV	subject object verb (word order)
SPL	superior parietal lobule
SVO	subject verb object (word order)
SWS	slow-wave sleep
T	thymine
ToM	theory of mind
TOP	temporal parietal occipital (junction)
TPJ	temporal parietal junction
U	uracil
UAL	unlimited associative learning
UCR	unconditioned response

UCS	unconditioned stimulus
V	verb (in a sentence)
VENs	von Economo neurons
VMPFC	ventromedial prefrontal cortex
VWFA	visual word from area
W	wakefulness

1

A Brief History of Life and Brain Evolution

Adaptations and Exaptations

Evolutionary neuropsychology is the study of the evolution of the structures and functions of the human brain. It is an exciting multidisciplinary science that spans the fields of evolution, neurocognitive sciences, psychology, anthropology, and many others. Importantly, it assumes that current human brain regions and their functions were adaptations to distant ancestral demands of varying conditions and environments, which now may serve different functions or have different purposes. Charles Darwin commented, in 1862, on the latter observation in one of his most obscure books, *On the Various Contrivances by Which British and Foreign Orchids Are Fertilized by Insects,* wherein he noted:

> Although an organ may have not been originally formed for some special purpose, if it now serves for this end, we are justified in saying that it is specially adapted for it. On the same principle, if a man were to make a machine for some special purpose, but were to use old wheels, springs, and pulleys, only slightly altered, the whole machine and all its parts, might be said to be specially contrived for its special purpose. Thus throughout nature almost every part of each living being has probably served, in a slightly modified condition, for diverse purposes, and has acted in the living machinery of many ancient and distinct specific forms. (p. 348)

An Evolution and Natural Selection Primer

At the outset of this book, it is important to define some of the terms of evolution that are often misunderstood. The theory of natural selection, as independently proposed by Englishmen Charles Darwin and Alfred Russel Wallace, states that changes in structure or behavior of a species across successive generations are due to their natural selection in their environments

Evolutionary Neuropsychology. Frederick L. Coolidge, Oxford University Press (2020).
© Oxford University Press.
DOI: 10.1093/oso/9780190940942.001.0001

because those traits or behaviors aid reproductive success or fitness (defined as the number of offspring left by an individual after one generation). Thus, "survival of the strongest" is a misconception about the nature of natural selection, and so is "survival of the fittest," unless the latter is defined as per Darwin and Wallace. It is not the strongest who survive but rather those who are more successful than others at reproduction in a particular environment; they are the ones who "survive" to pass their genetic makeup to succeeding generations. Therefore, natural selection is not a random process, but rather is based on random genetic mutations. These random mutations are an essential part of evolution, and they may result in positive or negative outcomes (and they may also have no demonstrable effect). Mutations with positive outcomes increase the reproductive success of an organism. The negative ones often result in spontaneous abortions, which obviously means that the fetus is taken out of the gene pool. Mutations with neither a positive nor negative effect (called *neutral mutations*) are probably common (thank goodness) but may or may not result in variations of structure or behavior (called *phenotypes*). It is the randomness of genetic mutations that has been very useful over evolutionary time, as variations in phenotype provide a kind of safety net when environmental conditions change drastically or quickly. If all organisms of one species had the same phenotypes, then, when environmental conditions changed suddenly, those organisms would be at risk of extinction (as the reader will see shortly with the common banana). So the process of evolution has an element of both randomness (as with genetic replication) and non-randomness, where natural selection favors the phenotypes resulting from genotypes (one's basic genetic makeup) that survive environmental challenges and reproduce more successfully than the phenotypes of other genotypes. Natural selection, therefore, does not select the strongest or "best," and natural selection has no long-term or final goal. In fact, in the final chapter, readers will see that natural selection does not even produce a best or perfect biological organism, as natural selection is gradual and constrained by previous selections, so organisms become adapted to their present environments, and when those environments change, natural selection lags behind. For example, when the ancestral environment favored those who were active, impulsive, and quick to act, those behaviors were favored (differential reproductive success). However, when the environments changed to favor slower activities (like agriculture) instead of hunting, or even semi-staid, mostly non-active present school environments, the genes that favored highly active individuals fell out of favor, and children with attention-deficit hyperactivity disorder (ADHD), for instance, were seen as having harmful or pathological behaviors. In summary, evolution may be viewed as an outcome. It is the result of random mutations in DNA replication and the subsequent differential reproductive success of the phenotypes of one's genotype and the flow of those genetic features across generations (see Al-Shawaf, Zreik, & Buss, 2018, for other misconceptions about natural selection).

About 120 years after Darwin and Wallace developed the theory of natural selection, paleontologists Stephen Jay Gould and Elisabeth Vrba (1982) coined the term *exaptation* to refer to features that have been co-opted from their initial adapted functions but which now enhance one's evolutionary fitness. They further proposed that the word *adaptation* be restricted, as Darwin had originally proposed, to features created by natural selection for their current purpose. A classic example of an original adaptation and later exaptation is feathers. There is convergent evidence that feathers were originally selected for the purpose of thermoregulation. Millions of years of genetic mutations later, lighter creatures had their reproductive fitness enhanced by the feathers because it afforded them greater mobility, flight, or both. This exaptation gave these creatures a reproductive fitness advantage over those who were slower and flightless. And so, too, it appears that different regions of the brain and even brain cells themselves have been selected over time to serve different purposes. One classic example in this book (see Chapter 5) is the parietal region. One of its original adaptations must have been the integration of sensory information to aid movements, either toward or away from other creatures and environmental conditions (foodstuff, dangers, etc.). Currently, there is mounting evidence that the parietal region is heavily involved in aspects of consciousness, including a sense of who we think we are and what we think of other people, and thus its exaptation appears to still be movement but in a metaphorical sense: who we are, what are we doing, and where are we going.

There is, certainly, danger in attempting to delineate what were the human brain's original adaptations or even its original functions. However, I find rather satisfying the idea that one of the brain's most important original purposes may have been movement. In this regard, computational brain models and computer models of brain structures and functions have always been rather unsatisfying to me. A computer's chief purpose or the reason for its creation has absolutely nothing to do with movement. Perhaps if we go back 600 million years to the first multicellular life forms, like comb jellies and sponges, then computer analogies might be relevant, as comb jellies had cilia for movement and juvenile sponges floated about until they became sessile. But I digress. Long ago, it appears that some cells were selected for their ability to categorize environmental conditions (sensory cells) and some cells aided movement (motor cells), and some cells specialized in coordinating those two activities for optimal functioning (like eating and reproducing). The latter cells became the first essence of modern brains.

Brain Cells

At the outset of this book, it is also important to discuss the two types of brain cells: neurons, of which there are estimated to be about 86 billion to 100 billion, and glial cells (about the same number as there are neurons). *Neurons* communicate with each other electrochemically, and they consist of an *axon*, which receives signals to fire from the cell body, sending an electrochemical signal along its length. *Dendrites* (branch-like structures), which are connected to axons of other neurons, receive this electrochemical message from its contact with the axons of 15,000 or more other neurons. These electrochemical signals are called *neurotransmitters*, of which there are many different kinds, such as dopamine, norepinephrine, serotonin, and others. The ends of the axons contain these neurotransmitters in synaptic vesicles (buttons or boutons), which, when signaled to fire, excrete their contents into the gap between the dendrites and the axons. This gap is called a *synapse* or *synaptic cleft*. Santiago Ramón y Cajal, an early 20th-century Spanish Nobel Laureate, once described synapses as "protoplasmic kisses" (Colón-Ramos, 2016). These adjacent dendrites have postsynaptic receptors, which receive their signals only from specific neurotransmitters and send that information to the cell body.

The other 50% of brain cells are called *glial cells* (which derives from the Greek word for "glue"), and they were originally thought only to support neurons nutritionally, such as providing oxygen and holding the neurons together. Recent studies, however, have shown that glial cells are far more sophisticated than originally imagined. Glial cells do seem to surround neurons and keep them in their place. They do provide oxygen and other nutrients to the neurons. They also protect neurons from pathogens, remove dead neurons, and clear synapses of unneeded chemical transmitters. Glial cells direct the migration of neurons to their appropriate place during brain development and appear to regulate the development of axons. They also regulate the growth of dendrites. There are at least four different kinds of glial cells, and one type, *astrocytes,* accounts for about 20 to 40% of all glial cells. They are important to the continued regeneration of the adult hippocampus (one of only two areas of the adult human brain that regenerates its cells; olfactory bulbs are the other). In fact, astrocytes have been likened to neural stem cells, which have an ability to change their original function while still being critical to all of the brain's developments. About 15% of all brain cancers are glioblastomas, which often involve astrocytes, and perhaps because of their stem cell–like nature, they are the most deadly form of brain cancer, with a strong a majority of victims dying within 3 to 15 months of diagnosis (sadly, this occurs with or without treatment).

Neural Reuse: A Central Organizing Principle of the Brain

Psychologist Michael Anderson (2010, 2014) has proposed that the reuse of brain tissue for purposes other than its original adaptation may not only be a pervasive feature of the brain's organization but also be considered a central organizing principle for the brain's evolution of higher cognitive functions. He labeled this idea the *massive redeployment hypothesis* (MDR). At its core, the MDR makes the assumption that "evolutionary considerations might often favor reusing existing components for new tasks over developing new circuits de novo" (p. 246). One of MDR's most important predictions, backed by substantial empirical evidence, is the notion that a typical brain region may support many diverse cognitive functions. Anderson gives the example of Broca's area, classically thought to be responsible for speech production, yet shown to be highly active in a variety of other cognitive and perceptual-motor tasks, such as the sequencing of actions, the preparation of motor movements, action imitation, the recognition of actions, and many others. In a review of 1,469 functional magnetic resonance imaging (fMRI) studies for a single cortical area, he found activation for nine different cognitive tasks. A necessary corollary of this prediction is that a single cognitive task would most often be controlled by multiple cortical regions.

There are many other neural reuse theories—for example, cognitive neuroscientist Stephane Dehaene (2005) has proposed the *neuronal recycling hypothesis.* Although it is similar to Anderson's massive redeployment hypothesis, it is more concerned with recently evolved cognitive functions like reading and mathematics rather than being concerned with a long evolutionary perspective for the reuse of brain cells. Cognitive functions like reading and mathematics may not have had enough evolutionary time to develop specific cortical circuits, but they do appear to have some specialized neuronal clusters. Thus, in order to learn to read or conduct mathematical operations, these functions must rely on pre-existing neuronal circuitry with some degree of neuroplasticity, which can support novel thoughts and novel procedures. For example, higher-level mathematical operations appear (in part) to depend on cortical regions in the medial temporal lobes, parietal lobes, the angular gyrus, and the intraparietal sulcus (all discussed in Chapter 3). The intraparietal sulcus has neurons dedicated to subitization (being able to differentiate between one, two, and three things) and conducting magnitude comparisons between smaller and larger sets (differentiating between 25 things and 50 things). These two core processes form a basic ability called *numerosity* (e.g., Coolidge & Overmann, 2012). Obviously, higher-level mathematical operations are also highly dependent on these neuronal circuitries, their accompanying neural plasticity, and the ability to engage other cortical areas in their operations.

A Brief History of Life

Life's origins can be traced to not long after the beginnings of earth itself. Earth began to accrete about 4.567 billion years ago, and living things began to appear about 3.9 billion years ago, about the time earth itself was less volatile from volcanic action and meteoritic bombardments. How life chemically and biologically began is still a bit of a mystery. That scientists have not yet duplicated the origins of life in a test tube does not necessarily imply supernatural origins. It may simply mean they have not yet discovered the exact conditions for the origins of life. One popular hypothesis is that there was a pool of just the right chemicals and a lightning strike provided the initial "spark." Other scientists posit that lightning is a relatively rare source of energy compared to a much more prominent and near constant source of energy, the sun. These scientists propose that it may have been the proper chemicals and sunlight which started life; however, this idea also has its detractors, as those chemicals and conditions have not yet produced life in a laboratory. Yet a third and newer hypothesis is that life came to earth in a meteorite or comet billions of years ago, and, while provocative, it still does not explain the processes by which life began, where it started, and how it was picked up by a comet. Nonetheless, scientists generally agree that what may be considered life began on earth about 3 to 4 billion years ago.

Perhaps a definition of life should be discussed before discussing its evolution. What are common characteristics of a living organism? First, living things need to take in chemicals or materials to sustain their energy and to expel the waste products of those energy reactions. This process is called *metabolism*. It is also possible that living organisms may produce their own energy, but that would still require an external source of energy and a means to convert it. A common example of such a process is photosynthesis, whereby a cell changes sunlight into chemical forms that can then provide energy for a cell's actions. Second, living things either need to reproduce or, once created, be immortal or semi-immortal (this hypothesis is not so outrageous, as some kinds of bacteria and viruses seem fairly hearty in terms of longevity, measured perhaps in billions of years).

A cell wall may have been the first evolutionary advance in life. The earliest of these walled-cells, called *prokaryotes*, had no nucleus, and a single strand of RNA (e.g., early bacteria). There were and are many variations of bacteria, which even now can live under extreme conditions (like in hot sulfur springs). Perhaps 1 billion to 1.5 billion years later some of these prokaryotic cells surrounded their RNA with another membrane to form a nucleus; they are called *eukaryotes*. Thus, prokaryotes can be considered the ancestors of all modern plant and animal life. It may be further argued that eukaryotes, with their nucleus directing their activities, may be considered a prototype of bodies and their brains.

The first groups of multicellular life, the comb jellies and sponges, emerged about 700 to 600 million years ago, and there was probably less oxygen in oceans at that time. Over that period, oceans increased in oxygen, partially, perhaps, through some of the feeding and excreting activities of prokaryotes, eukaryotes, and multicellular life. In order for life to sustain itself, it must be able to feed, excrete waste, and replicate or reproduce. All of those activities may have been enhanced by the ability to move. Organisms that developed cells designed for movement may have been at a distinct evolutionary feeding and reproductive advantage. Thus began the natural selection for cellular differentiation for movement (motor cells), for the detection of foodstuff, mates, and predators (sensory cells), and for cells that could coordinate the activities between those two kinds of cells (rudimentary brain cells).

Although greatly contested, it may be surmised that natural selection may not favor behavioral complexity per se, but it may favor diversification, as in times of environmental stress and disruption (e.g., volcanic explosions, meteoritic bombardments, etc.); "fitness" may be a function of a pre-existing trait, which allows survival in a disrupted environment (volcanic explosions, earthquakes, floods, etc.). A current example may be a particularly deadly banana fungus *(Fusarium wilt)*. Because most marketed bananas are of the same species *(Cavendish)* and susceptible to this virus, having a *variety* of species of marketed bananas, some of which might be immune to this fungus, would be an excellent idea.

Within about 20 million years or so of the comb jellies and sponges, or about 580 million years ago, there appeared the first animal life with clearly differentiated cell types (nerves and muscles), radial symmetry (its body centered around an axis and perhaps a mouth), and the ability to move. They were the cnidarians, and they had rudimentary eyes. About 550 million years ago, simple flatworms appeared. They are considered bilaterians—that is, if split down the middle, both sides are nearly identical, and it is thought that bilaterality aided movement. From comparative studies of fossil and living flatworms, it is suspected that these earliest forms of animal life had sensory cells for olfaction and vision (and perhaps the detection of movement and electrical fields of other life forms), motor cells guiding movements (toward and away), and specialized cells that could coordinate the sensory and motor cells. It is these latter coordinating cells that formed the basis for the first brains. Notice, too, that at this time rudimentary brains were already bilateral. It is not known why dual hemispheres were favored by natural selection, although it is speculated that it may have aided heat dissipation and that motor neurons controlling left and right sides of the body may have been more efficient and less disruptive if they were separated. Why, ultimately, the left hemisphere in humans controls the right side of the body, and vice versa, is also a mystery, although it has been speculated that it may have something to do with the turning behavior on the bottom of the ocean of a common ancestor of all vertebrates about 500 million years ago (e.g., de Lussanet & Osse, 2012, 2015).

Shortly after the evolution of flatworms, about 480 million years ago, eel-like ancestors appeared with cartilaginous backbones. They are known as the *chordates,* and chordate brains strongly heralded modern animal and human brains with a forebrain (the upper part of the brain), a midbrain (central brain structures like the hippocampus, and others), and the hindbrain, or lower brain, sitting on the upper spinal cord (structures critical to maintaining vital activities like breathing, blood pressure, heart rate, etc.). Conodonts (now extinct) are an example of one of these primitive chordates. They were eel-like with large eyes (at least in the sense there were cells that were light sensitive) and fins aiding their movement. A curious cartilaginous structure known as a notochord ran along its length. The animals' muscles attached to the notochord, and it appears to have allowed for flexible movement. A living relative of these ancient conodonts is the lancelet (*amphioxus*); they are still studied today for insights into early brain development, as lancelets do not seem to have a true brain but do have the three-part brain structure noted earlier.

About 480 million years ago, the first fishes appeared, with clearly delineated brains, although the forebrain, especially the upper surface (the cortex or cerebrum), was particularly small compared to the size of the fish. Even modern sharks have an exceptionally small cortex such that, if carefully removed, a shark can continue to swim and breathe for days, thus demonstrating that fishes, in general, are highly reflexive, although they can obviously learn through classical and operant conditioning (reinforcement or punishment). In the fossil record, the first coelacanth appeared about 410 million years ago. It was thought to be extinct until a specimen was caught off the coast of South Africa in 1938 and a second one was caught in the Comoro Islands in 1952. Brain autopsies of coelacanths, compared to those of other extant fishes, sharks, rays, frogs, and salamanders, revealed smaller olfactory bulbs but a well-developed visual system. The cerebrum (which includes the cortex and cerebral hemispheres) was smaller, with an average-size but bilobed midbrain, but the hindbrain (upper spinal cord structures) and cerebellum were larger. In summary, based on the assumption that extant coelacanth brains are similar to extinct coelacanth brains, more than simple rudimentary brains had evolved by 410 million years ago. Clearly, the ratio of the mass of their brains to their bodies was much smaller than in most modern animals, especially mammals and primates, but much like that in most modern fishes. Their brains are like those of modern amphibians, which may suggest an ancestral relationship to later amphibians (and, ultimately, to us). About 390 million years ago, the first tetrapod (four-limbed) fishes appeared, and these fishes used their fins as limbs to move about in shallow saltwater and freshwater. Some modern lungfish still resemble the fossils of the earliest tetrapodean fish.

The Essence of Cellular Differentiation and Brain Evolution

It is a bit misleading to note that with cnidarians and flatworms, there were suddenly clusters of cells resembling brains that processed sensory information (internal and environmental) from sensory cells and appropriately directed motor cells. The differentiation from prokaryotes to eukaryotes may be attributed to essentially one process, replication, and even the evolution from prokaryotes to eukaryotes took place over a billion or more years. In the act of life reproducing itself through DNA replication, there are variations, as it is not a perfect process, although it is amazingly accurate. In humans, for example, errors in DNA replication occur only about once in a billion replications. These errors are called *mutations*—most are probably harmless, some are harmful or deadly, and some are beneficial. If a mutation subsequently increases an organism's chances of reproducing (and its offspring's chances of reproducing), then those individuals pass their genetic information into the gene pool at greater frequencies than others. The latter process is the essence of Darwinian natural selection, and it is called *fitness*. As noted previously, in Darwin's study of orchids, he suspected that most modern organisms' parts may have had some previous function that was different from its present function. Thus, through adaptations and exaptations, brains began to evolve, and somehow in this long process, brains became more complex.

Back to Fishes

About 375 million years ago, a remarkable fish lived, which is now labeled *Tiktaalik*. It is remarkable because it reflects the beginning of a transition from sea life to terrestrial life. It was a fish whose forward fins had bones that heralded modern animals' shoulders, upper arms, elbows, and wrist joints and they were like the bones in the earliest terrestrial animals. It also had a flat head and a neck (modern fish, of course, have no real neck). The first of the true amphibians came not much later (in geological time), about 360 million years ago, and some, like *Acanthostega*, had gills and lungs. Reptiles made their first clear appearance about 300 million years ago, having evolved from these early amphibians. About 250 million years ago, reptiles diverged into two major groups: the Diapsids, which includes dinosaurs, modern reptiles, and birds, and the Synapsida, which includes modern mammals. Mammals have hair, middle ear bones, mammary glands (which give them their name), give birth to live young (with a few egg-laying exceptions like the platypus), and, most importantly, a much larger cortex than that in fishes, birds, or reptiles. The mammalian cortex appears so enlarged that it has gained the name *neocortex*, which appears to demonstrate that mammals were less reflexive, less instinctual, and more behaviorally flexible.

About 65 million years ago, a branch of mammals began their own speciation, the primates. Originally, they were small, nocturnal, insect-, leaf-, and fruit-eating, tree-dwelling creatures, and probably more socially oriented than any ancestor. As fruit is more nutritional and was necessary to fuel their proportionately larger brains, these early little primates undoubtedly had to compete with birds, reptiles, and other animals for fruit. It is surmised they did so by coordinating their foraging activities with vocal calls. Thus began the selection for social brains and the necessary apparatus, sensory and brainwise, for the production and comprehension of sounds, that is, the foundations for language. Evolutionary neuropsychologist Francisco Aboitiz and his colleagues (Aboitiz, 2017; Aboitiz, Aboitiz, & Garcia, 2010) have written about this highly specialized auditory/vocal neural circuitry, the phonological loop, and its importance in extending short-term verbal memory and increasing the capacity to not only hold and process sounds but also set the foundation for more complex utterances and communications. They also proposed that the phonological loop became a mechanism for transferring sounds to long-term memory, and they traced its origins to these early primates.

About 40 million years ago, the primates diverged into two lineages, the simians (monkeys, apes, and us) and the prosimians (lemurs and lorises). About 20 to 15 million years ago, the taxonomic family Hominidae appears, which includes the great apes (also called *hominids*)—gorillas, orangutans, chimpanzees, and the bonobos (pygmy chimps)—all extinct ancestors of *Homo sapiens* (the first term is the genus, the second is the species), and us. As there is some confusion in the literature about the terms *hominid* and *hominin*, I will refer to *hominids* as the group consisting of all modern and extinct great apes, and *hominins* will be the more narrow term referring to all extinct and extant humans, and all of our immediate and distant ancestors (after the chimpanzee line diverged from ours, about 6 to 13 million years ago [that dating has also become a recent bone of contention, and the divergence between chimps and ancestors of *Homo sapiens* may have occurred more than once]).

Early Hominins

The fossil record for early hominin evolution, beginning about 5 million years ago or earlier, is a bit spotty, as hominins appeared to evolve primarily in moist, jungle-like conditions that are not conducive to making fossils or preserving well. One hominin who made a media splash in the 1990s was *Ardipithecus ramidus* (*Ardi* means "ground" or "floor," *pithecus* means "ape," and *ramid* means "root"). At first it was thought to be one of our more distant ancestors, but now it is thought to be an extinct cousin of modern *Homo sapiens*. It dates to about 4.4 million years ago, with a brain size of about 300 to 350 cc (the brain size of modern chimpanzees is about 390 cc, with a range from 275 to 500 cc). So the brain of *Ardipithecus ramidus* appears to be what would normally be expected of an early primate, that is, small brains relative to their body mass and a reliance on visual and auditory systems. The initial divergence from other African apes included a transition to partial terrestrial bipedalism (walking on two legs), yet the first hominids were essentially bipedal apes. Combined with evidence for an ape-like maturation rate, and lack of evidence for stone tools or other "new" cultural behaviors, there is no reason to suppose that these early ancestors of *Homo sapiens* were cognitively much different from current African apes. However, there would be a more dramatic change in hominin evolution about a million years after the initial appearance of *Ardipithecus ramidus*. The latter would be our earliest definitively known ancestor, *Australopithecus afarensis* (literally, southern ape from the Afar triangle in Eastern Africa), or better known as Lucy (see Fig. 1.1). It is currently thought that *Ardipithecus ramidus* might have been a "cousin" of Lucy but likely not our direct ancestor.

Lucy in the Sky with Diamonds: Our Most Suspected Distant Ancestor

In 1974, paleoanthropologist Donald Johanson and his colleagues were digging in a remote section of Ethiopia called the Afar Triangle, and just before lunch one day he spied a fossil elbow joint, a thigh bone, the back of a skull, some vertebrae, and a pelvis. He recognized that all of the bones were from a female individual and that they were much older than any other hominid fossil find to date. As the Beatles' song, Lucy in the Sky with Diamonds, was playing in the camp, Johanson nicknamed her "Lucy" (Johanson, Johanson,

Fig. 1.1 Lucy (*Australopithecus afarensis*) (See color plate).
Source: Wiki Commons.

Edgar, & Blake, 1996). Her official name is *Australopithecus afarensis*. Many australopithecines have been found since then, and they date to about 3.9 million years to about 2.9 million years ago. Their brain sizes range from about 375 to 550 cc (our modern brains average about 1350 cc). Lucy's skull is similar to that of a chimpanzee, except for having more human-like teeth. Lucy had an ape-like face with a low forehead and a bony brow ridge (called a *supraorbital torus*, which persisted in hominins even in Neandertals; however, it may have little or no explicit function and may have arisen through simple genetic drift [changes due to chance]). Male australopithecines (about 4 ft 11 in. [151 cm], 92 lb [42 kg]) were larger than females (about 3 ft 5 in. [105 cm], 64 lb [29 kg]). This males-larger-than-females condition is known as *sexual dimorphism*, which implies that male australopithecines may have fought among each other for reproductive access to females. Australopithecines have curved finger and toe bones, and their arms and legs were proportionately longer than those of modern humans. This skeletal evidence suggests that the australopithecines were bipedal (unlike modern chimps, who knuckle-walk, which is less metabolically efficient), but they probably slept in nests in trees at night for safety and played in trees, and females nested their babies in trees. Tooth analyses have revealed that the australopithecines ate soft foods like plants and fruit and some hard foods like USOs (underground storage organs, like roots), and undoubtedly they supplemented their diets with scavenged meat—"undoubtedly" because brains are expensive metabolic tissue, accounting for about 2% of a human body's mass but requiring about 20 to 25% of one's total calories. With a life in and among trees and in nests, they probably did not use or even manage fire.

About 3.4 million years ago, the first stone tools appeared (McPherron et al., 2010). They were found in the Dikika area of Ethiopia. They were mostly sharp flakes used for scrapping meat from scavenged (most likely) bones. The flakes were made (knapped) from larger stones (called *cores*). The cores could also be used as hammer stones to crack open nuts and break open bones for their marrow. These flakes and cores are called Mode 1 stone tools (Fig. 1.2). Later, some partial australopithecine skeletons were found in the same area, so paleoanthropologists believe that the knappers of these stone tools were the australopithecines.

I'll Be Your Handyman: *Homo habilis*

About 2.5 million years ago, a hominin appeared in the anthropological record that was physically similar to the australopithecines, about 4 ft tall (100 to 135 cm) and about 70 lb (32 kg). Its first skeleton was discovered by paleoanthropologists Louis and Mary Leakey in the Olduvai Gorge, Tanzania (Leakey & Leakey, 1964). The site was unequivocally associated with Mode 1 stone tools and evidence for butchering, which meant that meat remained a part of this hominin's diet. But the biggest surprise was its braincase. At 640 cc (range for all specimens, 500 to 690 cc), its brain was over 50% larger than the average australopithecine. A recent study (Spoor et al., 2015) estimated the Leakey specimen's brain size to be even larger, at 729 to 824 cc, which meant its brain size nearly doubled compared to that of australopithecines. Because at that time (1964) no stone tools had yet been found associated

Fig. 1.2 Mode 1 Oldowan stone tool (upper) and flakes (lower) (See color plate).
Source: Frederick L. Coolidge.

with the australopithecines and because brain size had increased dramatically, Leakey and his colleagues decided to place this specimen in the genus *Homo* and gave it the species name *habilis* (*handy* in Latin). However, because of its small body and long limbs (similar to Lucy), it was still suspected of living and sleeping in trees. Again, because brain tissue is so metabolically expensive, *Homo habilis* must have increased the percentage of meat in its diet. And it is known that meat contains the critical amino acid phenylaline, which converts to tyrosine, which ultimately converts to the important chemical neurotransmitter dopamine (e.g., DeLouize, Coolidge, & Wynn, 2016). There are arguments that *Homo habilis* should not have been assigned to the genus *Homo* but remain in the genus *Australopithecus*; however, given the near doubling (or at the very least a 50% increase) of its brain size and the consequence of a greater level of neurotransmitters like dopamine in the brains of *Homo habilis*, a jump to the genus *Homo* does not seem unwarranted. The skeletons of *Homo habilis* have also been found to date as recently as about 1.6 million year ago (making them unlikely to be the direct ancestor of *Homo erectus*, to be discussed next). There is some variation of morphology in the *Homo habilis* group, so they are sometimes referred to as the *habilines*. This variation also seems true of *Australopithecus*, so they are referred to as the *australopithecines*.

One Tall Adolescent: Nariokotome

In 1984, a nearly complete skeleton (except for feet and hand bones) was found near a region called Nariokotome, near Lake Turkana, Kenya. It was identified as a boy who died when he was about 8 to 13 years old (recent estimates tend toward 8 or 9 years old) and he lived about 1.6 million year ago. He was nicknamed Nariokotome (alternatively, Turkana Boy), and once again the anthropological world went topsy-turvy with his discovery. First, he was not physically like the australopithecines or habilines. He was obviously much taller, about 5 ft 3 in. tall (1.5 m) and weighed about 150 lb (68 kg). He probably matured more in an ape-like pattern than a modern human (or somewhere in between), thus, he might have experienced a "faster speed of life," which meant his teeth matured more quickly, he would have had a shorter adolescent period, and he would have attained an adult height and weight earlier than modern humans. It was recently estimated (Ruff & Burgess, 2015) that he would have attained an adult height of about 5 ft 9 in. (176 to 180 cm) and weighed about 176 to 183 lb (80 to 83 kg). Clearly, hominins this tall and heavy could not have lived in trees, and anthropologists recognized this change by classifying such types as *Homo erectus* (upright man). Again, there are many variations in the species, so the group is sometimes called *Homo erectus sensu lato* (in a broad sense). It has also been proposed that *Homo erectus* morphology may have been selected for long-distance running (Lieberman, Raichlen, Pontzer, Bramble, & Cutright-Smith, 2006). Nariokotome's brain size was also remarkable, at 880 cc, and it was estimated to have an adult capacity of about 910 cc, the species averages being about 930 cc (750 to 1,250 cc). As modern humans' average brain size is about 1,350 cc, *Homo erectus*' brain was about 70% the size of modern humans, but not the same brain shape, as will be shown later.

There were also some remarkable archaeological finds associated with *Homo erectus*. Symmetrical, bifaced, leaf-shaped stone tools (also called *handaxes*) were found that were so different from and seemingly technically advanced than the previous Mode 1 stone tools that they were labeled Mode 2 stone tools (Fig. 1.3). Their design apparently was so practical and useful that it persisted for well over the next 1 million years.

Fig. 1.3 Mode 2 Acheulean handaxe (See color plate).
Source: Frederick L. Coolidge.

Life on the Ground

Australopithecines and habilines lived most of their lives in trees, primarily for protection. However, there was probably a limit to the number of nests in a single tree because if there were too many nests, it might have the opposite effect and attract predators. Because of their size, *Homo erectus* had to have made the complete transition to ground life, and with that transition came a suite of new challenges, the first of which was predation. It is thought *Homo erectus* may have survived, in part, by strength in numbers. Anthropologist Robin Dunbar (1998) believes that tribes of *Homo erectus* and later hominins may have approached 150, while, if extrapolating from modern chimpanzee troop size to the australopithecines and habilines, individual families might have numbered 3 to 6 members, and a community might have averaged about 40 to 50 members. Dunbar's (2013) social brain hypothesis proposes that bigger brains were selected for in order to keep track of a larger number of individuals in a group—that is, who helps, who cheats, who is lazy. As social hierarchies are ubiquitous to almost all extant nonhuman primates, a socially oriented brain would have been necessary for these larger groups of *Homo erectus* for efficient and successful formation of alliances, recognition of alpha members, and other social tasks.

The increase in brain size in *Homo erectus* again meant an even greater reliance on meat, and tooth analyses of *erectus* do reveal a greater reliance on meat, but also on plants and other foodstuff. Another important technological development at this time, around 1.5 million years ago, was the use of fire. If, as some anthropologists think, *Homo erectus* was able to manage the effects of fire (or even being able to create it intentionally), then there may have been a cascade of cultural advancements. First, fire could have been used as added protection against predators (people or animals). Even many modern-time hunters and gatherers use fire at night as protection when hunting. Second, the light from fire could extend *Homo erectus'* day, allowing activities such as stone knapping to continue after dark. Third, fire could increase the nutritional value of food, make hard, inedible foods edible and more digestible, reduce the amount of chewing required (and time spent chewing), kill bacteria in plants and animal foods, and make food taste better. Fourth, fire could have been used to harden the points of spears and change the nature of some types of stone to make it more knappable. Fifth, as psychologist Matt Rossano (2010) has proposed, sitting around a fire may help bond a group socially. And if *Homo erectus* had some type of protolanguage, stories and myths could have been told around the fire, which might have enhanced the fire-bonding experience. Paleoanthropologist Jean Luis Arsuaga (2002) has proposed that story- and myth-telling might have been one of the subtle cognitive differences that made *Homo sapiens* more successful than any of their extinct ancestors or cousins (e.g., Neandertals).

The Weed Species

As the number of individuals in a group expanded, so too would the size of the territory necessary to sustain them. It is thought that the territory of *Homo erectus* may have expanded 10-fold over that of the australopithecines and habilines, to perhaps over 100 square miles (about 2.6 million square meters). Wells and Stock (2007) have hypothesized that *Homo erectus* represented the first of the hominins to achieve the epitome of fully metabolically efficient bipedalism (including a body designed for long-distance running), which would have aided territorial expansion. Interestingly, it has also been suggested that the development of inner ear bones in *Homo erectus* may have represented improved balance and coordination. Full efficient bipedalism would also have allowed *erectus* to explore new territory, and the challenges associated with new territory may have helped select for behavioral plasticity (i.e., an ability to solve novel problems) that further challenged and enhanced the cognitive abilities of these bigger-brained hominins. It is also known that beginning at least 1.8 million years ago (and probably earlier than that), *Homo erectus* began leaving Africa, as fossils of a type of *Homo erectus* have been found dating to 1.81 million years ago in Dmanisi (Georgia), Russia, and to about 1.4 million years ago in Atapeurca, Spain. The fossil remains of *Homo erectus* dating to at least 700,000 years ago have been found in Indonesia, but more amazing is that *erectus* may have persisted living in that region until about 70,000 years ago or even more recently. The discovery of *Homo floresiensis* (found on the island of Flores, Indonesia) "hobbit-like" fossils with very small brains (about 400 to 500 cc) suggests that it was the australopithecines who may have left Africa first. It has also been suggested that they were a lineage of *Homo erectus*, already in Asia, and that they adapted to a restricted island environment by being smaller bodied and brained. Cachel and Harris (1995) were the first to suggest that *Homo erectus* may be likened to a weed species, able to invade distant, disrupted environments and to thrive in them. Wells and

Stock (2007) proposed that this behavior may have actually begun with the australopithecines but reached a demonstrable zenith with *Homo erectus*. As they further note, this capacity to adjust to fluctuating and unpredictable environments favors behavioral plasticity that ultimately protects genetic variation (always a good thing when environments change suddenly or drastically) and favors its release under environmental stress. I shall address the cognitive and neurological underpinnings for *Homo erectus'* behavioral plasticity in later chapters.

A Grade Change

Anthropologists refer to a "grade" in cladistics (the classification of animals and plants) when a related group of species shares similar behaviors and other characteristics. They refer to a grade shift when a group shares distinctively new behaviors and adaptations. Most anthropologists do not think the early hominins such as the australopithecines and habilines represent a grade shift different from the ancestors of orangutans, gorillas, and chimps. Even though there may have been the inchoate beginnings of a grade shift with the australopithecines, as argued by Wells and Stock (2007) with reference to the australopithecine radiation into more variable environmental niches, nearly all anthropologists agree that *Homo erectus* represents a grade shift from an ape-like way of life to a more human-like pattern of living.

The Next Grade Shift?

In 1908, near Heidelberg, Germany, a fossil was found that seemed to be an intermediate between *Homo erectus* and the early ancestors of modern humans. It was recognized at that time as perhaps a different species, and it was named *Homo heidelbergensis*. These individuals were stockier and shorter than *Homo erectus,* with males being about 5 ft 9 in. (175 cm) and 140 lb (64 kg) and females about 5 ft 2 in. (157 cm) and 115 lb (52 kg). Their physiology suggests that they had become cold-adapted, unlike *Homo erectus,* who appeared to be heat-adapted. It is not well established whether *Homo heidelbergensis* evolved in Africa and moved northward or whether *heidelbergensis* evolved from *erectus* in Europe or Asia, became cold-adapted, and moved back to Africa. However, it is thought that *Homo heidelbergensis* was the common ancestor of both modern *Homo sapiens* and Neandertals (and perhaps the Denisovans and others). Nevertheless, about 1 million years ago (plus or minus a couple of hundreds of thousands of years), a new species appeared with another dramatic increase in brain size, whose range overlaps with modern human brain size, from about 1,100 to 1,400 cc. It is important to note, however, that their brains overlapped with modern human brain size, but not with modern brain shape, as will be discussed in greater detail later. There is some controversy as to whether *Homo heidelbergensis* was the first cold-adapted hominin, as some researchers have proposed an earlier species, *Homo antecessor,* appearing about 1.2 million years ago. However, until more fossil evidence for *antecessor* is found and genetic analyses are conducted, it might be more conservative to assume *antecessor* was simply an earlier transition from *erectus* to *heidelbergensis.*

Our Cousins, Neandertals, Denisovans, and Others

The European version of *Homo heidelbergensis* gave rise to at least one other lineage, the Neandertals (it may also be spelled *Neandertal,* an original German spelling, but in German it is always pronounced *Neander-tall*). The African version of *Homo heidelbergensis* may have given rise to our lineage, *Homo sapiens.* Although Neandertals and modern humans have a high degree of DNA similarity (about 99.5%), it is the very small differences in our DNA that may have led to Neandertal's extinction and our evolutionary survival. The two groups had separate evolutionary histories for about half-a-million years; Mendez, Poznik, Castellano, and Bustamante (2016) have dated the lineage split to about 588,000 years ago. DNA analyses also indicate that some gene flow occurred between Neandertals and some modern humans. Modern Eurasians have been found to have 0.5 to 4% of DNA from interbreeding with Neandertals about 80,000 to 50,000 years ago, although modern sub-Saharan Africans usually have no DNA from Neandertals (Green et al., 2010). Mendez and his colleagues also found specific protein-coding genetic differences between Neandertals and *Homo sapiens,* all of which had potential damaging effects on the viability of any hybrids. It also has been proposed that male hybrids were particularly infertile, owing to antigens that may have been elicited as a maternal immune response to male hybrid fetuses during pregnancy. Juric, Aeschbacher, and Coop (2016) have suggested another factor, which ultimately increased the reproductive isolation between the two groups and may have involved some of the deleterious genes that *Homo sapiens* inherited from Neandertals. They proposed that most of the alleles (alternative forms of a gene) were essentially neutral in Neandertals but were selected against in the larger-sized populations of *Homo sapiens*. Although the alleles individually probably had small effects, these alleles may have posed a greater threat to the early generations of *Homo sapiens*–Neandertal hybrids. Ultimately, the last Neandertal died about 30,000 years ago.

Earlier hominins, such as *Homo erectus*, abandoned Europe during glacial cycles, but Neandertals and their immediate ancestors stayed and adapted, except perhaps for the brief periods of extreme cold. Many of the derived (unique) features of their anatomies reflect this local adaptation. At the outset, it is important to note that when taken individually, most Neandertal characteristics fall within the range of variability of *Homo sapiens*, modern and ancient; it is the combination of features that allows us to distinguish them.

Who Were the Neandertals?

The first partial skeleton of a Neandertal was discovered in 1829 in Belgium, but it was not identified then as such. German physician Johann Fuhlrott recognized some bones found by quarry workers in the Neandertal Valley, Germany, as human but different from contemporary humans, and Fuhlrott and a professor of anatomy, Hermann Schaaffhausen (1865), published Fuhlrott's findings in 1865. In 1863, the bones were labeled as a new species of human, *Homo neandertalensis*. By modern human standards, Neandertals were relatively short, even perhaps slightly shorter than their ancestor *Homo heidelbergensis*. Neandertals were stocky and powerful. The average males' height was about 5 ft 6 in. (165 cm) and they weighed about 171 lb (78 kg). Female Neandertals were about 5 ft 1 in. (156 cm) and weighed about 146 lb (66 kg). They had a round, barrel-like trunk. The muscle attachments on their bones indicated that they were heavily muscled. Their forearms and lower legs were relatively short compared to their upper limbs, and their forearms may have been slightly bowed. It is the classic build of a modern bodybuilder, similar for male and female Neandertals, and these features appeared early in their development. They also appeared to have led strenuous lives, and these physical characteristics were part of their genetic makeup.

Neandertal Skulls and Brains

The Neandertal skull was a combination of ancestral characteristics, cold adaptations, and derived features perhaps attributable to simple genetic drift (like their big, overhanging brow ridges). Their ancestral characteristics included a bigger, longer, relatively flatter skull than *Homo sapiens*, with large faces, teeth, noses (for heat exchange during breathing), eye sockets, and expanded frontal sinuses (most of these features reflect cold-weather adaptations). The one feature that would seem to be a dramatic exception to this similarity to *Homo heidelbergensis* was brain size.

Neandertal brains were about 9 to 13% larger than those of modern *Homo sapiens*. Estimates of their cranial capacity range from about 1,250 to 1,740 cc, with mean cranial capacity varying from 1,430 to 1,550 cc (depending on the sample size and ratio of males to females in the sample). However, a critically important notion in brain evolution is that at some point bigger brains stopped being associated with greater intelligence, or more cognitive flexibility, or greater adaptability. Neandertal brains were also much larger than *Homo erectus* brains, but they were not simply scaled-up versions of *erectus* brains. Neandertal brains had a different shape than that of *erectus* or modern human brains. After a review of brain structures, networks, and functions in Chapter 3, I will summarize the brain shape differences and their potential functional significance between Neandertals and *Homo sapiens* based on the discoveries by paleoneurologists Emiliano Bruner, Markus Bastir, Phillip Gunz, Simon Neubauer, and others.

There is another critical difference between these two human types that appeared after birth: a brain globularization phase unique to *Homo sapiens*. It occurred during the first year of life (Gunz et al., 2012), and it resulted in a "rounder" yet smaller brain in *Homo sapiens*, which appears to result primarily from a parietal lobe expansion (e.g., Bruner, 2010; Bruner & Iriki, 2016). There is some evidence that Neandertals had slightly shorter life spans, compared to *Homo sapiens* living at the same time, perhaps due to a more dangerous life style, nutritional insufficiencies, disease, xenophobia, or even a penchant for cannibalism (e.g., Agustí & Rubio-Campillo, 2016); evidence for the latter speculations will be addressed later.

The Denisovans and the Unknown Hominin

At Denisova Cave in southern Siberia, archaeologists found fragmentary skeletal remains, which, given the stone tools found with them, they assumed came from Neandertals. At first, only the DNA from a single fragment of a finger bone was analyzed, and they determined that it came from a previously unknown hominin who had died about 41,000 years earlier. The DNA analysis revealed that this lineage of hominins, subsequently named the Denisovans, shared a common ancestor with Neandertals, who had lived in Southeast Asia, Siberia, and even Western Europe. The Denisovan and Neandertal lineages appeared to have diverged from each other about 390,000 years ago. Denisovans also lived with the ancestors of some modern humans in Melanesia and Australia, and about 3 to 5% of the DNA of current residents in those areas is derived from the Denisovans. Other DNA studies showed that Denisovans also interbred with Neandertals in Western Europe, probably frequently, and recently a genome analysis revealed that a Neandertal mother and Denisovan father successfully interbred about 50,000 years ago (Slon et al., 2018). Most amazingly, the scientists detected the distinctive DNA of an unidentified ancient hominin. One interesting thing about that finding is anthropologists have yet to find any bones of that hominin, and the only evidence for its existence comes from DNA analyses (Sawyer et al., 2015). Preliminarily, though, it appears that this hominin diverged from the lineage that led to modern humans, Neandertals, and Denisovans. It may have had a common ancestor with *Homo heidelbergensis* or it may have been a branch of the *heidelbergensis* lineage (or with some other lineage of *Homo erectus*). However, with such sparse evidence to date, it is obvious that many discoveries lie ahead and many wonderful mysteries remain.

Why Only Us?

One of the major mysteries in anthropology is why *Homo sapiens* survived and Neandertals and other hominins went extinct. With such little anthropological evidence for the Denisovans and other hominins, speculation has mostly centered on the Neandertal extinction, and it is a surprisingly contentious debate. Anatomically modern *Homo sapiens* left Africa for Europe at least 100,000 years ago. They made it to the Levant (roughly the regions of

Israel, Turkey, Jordan, Lebanon, Syria, and others), and apparently retreated presumably when they encountered Neandertals who had already been living there for tens of thousands of years. Sometime around 45,000 years ago, *Homo sapiens* swept through the Levant and into Europe. It appears they were anatomically modern (skulls and brains and bodies) and carried with them a complex culture known as the Aurignacian, which included sophisticated cave art, creative figurines, pervasive use of personal ornaments, and highly ritualized burials. Within 15,000 years of their entry into Europe and Asia, Neandertals went extinct. There is little or no evidence of direct violence between the two human types. Currently, there are two general and opposing lines of thinking about Neandertal extinction. One is that the two human types were identical culturally and, therefore, cognitively. Although not explicitly stated, this side generally endorses the proposition that "there but for the grace of God, go us." In other words, it was a simple twist of fate, and if history replayed itself, *Homo sapiens* would go extinct instead of Neandertals about half of the time. The other position, often vilified with calls of racism and Victorian values (e.g., Villa & Roebroeks, 2014; Zilhão, 2014), is that a small but significant difference in cognition, probably due to brain shape differences, allowed *Homo sapiens* to extract more resources from the same environments and not only survive but flourish (e.g., Coolidge & Wynn, 2016, 2018; Wynn & Coolidge, 2007b, 2010, 2012, 2016; see Wynn, Overmann, & Coolidge, 2016 for a summary of these arguments). Because there is strong evidence that Neandertal groups had fewer members than *Homo sapiens,* any tiny difference in birth and death ratios would have hastened their demise in the face of competition, from other Neandertals and from *Homo sapiens.* As this evolution of brain shape differences is crucial to this book, it will be addressed again later.

Summary

1. An *adaptation* is a physical or behavioral feature which, through natural selection, aids survival and reproduction.
2. *Exaptations* are physical or behavioral features that have been co-opted from their initial adaptive functions and subsequently enhanced fitness.
3. The reuse of brain neurons for purposes other than their original adaption may be considered a central organizing principle of the brain.

4. The term *hominin* refers to all current and extinct relatives and ancestors of *Homo sapiens*, including the australopithecines and habilines, within about the last 6 million years.

5. *Homo sapiens* survived and flourished, instead of Neandertals, Denisovans, and other hominins, because of brain shape differences, which created cognitive differences that enhanced evolutionary fitness.

2

The Evolution of Learning and Memory Systems

None of the behavioral or structural phenotypes discussed throughout this book, whether acquired as adaptations or reused as exaptations, would be possible without processes that allow learning and memory. So, at the outset, I will present some basic definitions for learning and memory. *Learning* can be defined as a response, knowledge, or skill that is acquired by instruction, practice, experience, or study. Linguists Sukhoverhov and Fowler (2015) noted that memory can be referred to either literally or metaphorically. If, as they claim, one definition of memory is that it is "a property of matter such that it maintains its states (structures) for short or long periods of time" (p. 49), then rocks and water have memory. Indeed, molecular biologist Norbert Hertkorn and his colleagues (2013) study the biogeochemical signatures of oceans, and they have found these signatures to be stable and vary consistently by ocean depth and location. These chemical signatures can be analogically considered "memory"; however, I will confine memory definitions more literally to living biological organisms (not oceans). Cognitive psychologist Endel Tulving (1995) defined memory thusly: "Memory in biological systems always entails learning (the acquisition of information) and . . . learning implies retention (memory) of such information" (p. 751).

Learning and memory have played a prominent role not only in science, particularly psychology and the cognitive sciences, but also in literature and the movies. They range from humorous observations by Mark Twain, such as "A clear conscience is the sure sign of a bad memory," and "If you tell the truth, you don't have to remember anything," to more profound (confounding) observations, such as "Right now I'm having amnesia and déjà vu at the same time. I think I've forgotten this before," by comic Steven Wright. One of the more powerful movies that seriously questions who individuals are without their memories is *Memento*, where the hero of the film is presumably stabbed in the hippocampus and loses track of his and others'

Evolutionary Neuropsychology. Frederick L. Coolidge, Oxford University Press (2020).
© Oxford University Press.
DOI: 10.1093/oso/9780190940942.001.0001

identities. In his struggle to find out who his friends and enemies are, he takes Polaroid photos and makes notes on the backs of them so he can remember whom to trust and whom not to trust. Although the cause of his memory loss in the film is neurophysiologically questionable, the dire consequences of severe memory loss are profound. Further, those who live long enough to become demented (loss of reasoning and memory) not only often become unable to take care of themselves, but their memory losses can have a major psychological impact on their caretakers, especially when a demented father or mother asks their adult child, "Who are you?"

This chapter, however, will not focus on memory loss but instead on some typical categories of learning and memory, and there will be a discussion of their evolutionary origins and molecular basis. I shall begin with one of the most common learning dichotomies: associative and non-associative learning. If there is a *memory* for some "thing," there has to have been initial learning of that "thing" (I say "initial" because the "thing" could have been passed on to subsequent generations by genetic memory). Further, distinctions are often made between short-term and long-term memory, although the dichotomy is a bit arbitrary; *short-term memory* is the recall of learned material over short periods of time, like seconds or minutes, while *long-term memory* is the recall of material over hours, days, or years. There is yet another prominent dichotomy for memory: explicit versus implicit memory. *Explicit memory* is the recall for facts, words, meanings, and events. *Implicit memory* is the recall for skills or procedures, like riding a bicycle, juggling, or knapping a stone into a handaxe. It is also important to point out that the relationship between the learning and memory dichotomies is murky and rarely addressed, but I shall attempt to elaborate on one here. I will end the chapter with a discussion of a multicomponent theory of learning and memory, *working memory*, as originally proposed by experimental psychologist Alan Baddeley in 1974 (with his postdoctoral student, Graham Hitch).

Non-Associative and Associative Learning

Most presentations of associative and non-associative learning place associative learning first (alphabetical, more alliterative, tradition, etc.). I have intentionally reversed the standard order in my subtitle because non-associative learning may have preceded associative learning by a billion years or more. First, however, a definition: *Non-associative learning* involves changing the strength of a relationship between a stimulus and a response, based on repetition of a stimulus. The two forms of non-associative learning are habituation and sensitization. In habituation, the strength or probability of a response is reduced by repeated presentation of a stimulus. *Habituation,* in essence, means learning not to respond to irrelevant stimuli. It is arguably the simplest form of learning, as it has been observed to occur in virtually all species of animals and some plants (for the latter, see Gagliano, Abramson, & Depczynski, 2018). In the traditional definition of sensitization, the strength of a response is increased by repeated presentation of a stimulus. However, I see this definition as being too narrow. I argue that sensitization should include steady or reliable responses to repetitive stimuli. Thus, *sensitization* means that the organism learns to respond to relevant stimuli.

It's the Beginning of the World, and We Don't Know It

I give my apologies to the band R.E.M. for the subtitle of this section, although it is kind of them to lend their name to such an important stage of sleep. Spanish molecular biologist Victor de Lorenzo (2014) has argued that the central dogma of molecular biology, "DNA [deoxyribonucleic acid] makes RNA [ribonucleic acid] makes protein" (p. 226), ignores the importance of *selfish metabolism.* He hypothesized that selfish metabolism is the main evolutionary drive of DNA-containing organisms. He proposed that it is not replication and expansion of the DNA in organisms that is paramount but the exploration and exploitation of organisms' chemical landscapes. Locating and subsequently exploiting metabolism-enhancing chemicals helps to ensure organisms' enhancement of their metabolic processes, which *subsequently* helps to ensure their replication and introduction of their DNA in the gene pool. However, a paradox arises: Molecular biology's focus on DNA appears to have a flaw, at least from an evolutionary perspective. DNA may be considered a modified version of two strands of RNA, as it is unlikely that RNA and DNA evolved separately. Thus, RNA must have preceded DNA in evolutionary time. The latter idea is known as the *RNA world hypothesis—* that is, in prokaryotes (cells without a nucleus), RNA alone stored both genetic information and catalyzed chemical reactions. Interestingly, fossil evidence shows that prokaryotes appeared about 3.9 billion years ago, and the earth formed about a half a billion years earlier. Further, RNA has persisted in eukaryotes (cells with a nucleus), so the idea that RNA may be likened to "fossils" of the origins of life is intriguing. I also find the appearance of prokaryotes 3.9 billion years ago very surprising, so relatively recent after the earth's formation. These incredibly complex prokaryotes had already evolved cell walls, a membrane surrounding cytoplasm, RNA, chemical receptors (to detect beneficial and harmful chemicals) and mechanical receptors (to detect obstacles), and flagella that guided movements toward and away from chemicals and objects. So, how did all that evolution of cells and its materials take place in such a relatively short period of time? One highly provocative idea is that prokaryotes did not evolve—on earth; that is, they came from a meteorite or comet. Of course, that still leaves the question of how these complex life forms developed and how they got inside a meteorite or comet. And what were the initial conditions for their evolution, wherever they evolved? Nonetheless, given that these molecules and initial life forms had undoubtedly shorter lifespans compared to later and more complex life forms, perhaps a half a billion years may have been sufficient for their evolution on earth.

Prokaryotes Show Non-Associative Learning and Short-Term Memory

The modern model for the study of prokaryotes is the bacteria *Escherichia coli* (*E. coli*), which is one of the mostly healthy microbiota in warm-blooded animals, including humans. The prokaryote cell is (and *E. coli*) enclosed in a cell membrane, which houses cytoplasm, RNA, and unwalled organelles (chemical clusters). Thousands of species of prokaryotes have been identified, and many more are suspected. On the basis of *E. coli* studies, their stimuli recognition and information processing depends on previously established learning and memory, which are contained within the prokaryotes' RNA. These learning and memory systems can help them identify food sources, identify adverse events, avoid and protect against toxins, and avoid and bypass obstacles. *E. coli*'s chemoreceptors are particularly sensitive to glucose and amino acids such as aspartate, which helps synthesize proteins. Their non-associative learning may work like this: If attracted to a beneficial chemical (sensitization), the cell's flagella are coordinated, which moves the cell toward the chemical. If repelled by the chemical, if the chemical is irrelevant, or if the chemical concentration drops, the flagella become uncoordinated, and the cell tumbles away from the chemical. I consider this an example of habituation. Apparently, the chemoreceptors continually detect the concentration gradient over periods of about 3 seconds, and the cell "remembers" the gradient changes and responds with its flagella accordingly. *E. coli* does not possess any true visuospatial abilities, but the short-term memory they have is enough to orient it appropriately to its chemical landscape, that is, to habituate to irrelevant chemical stimuli and be sensitive to relevant chemical stimuli. It may also be assumed that a homeostatic balance threshold between habituation and sensitization was naturally selected, which varied among individuals between hypersensitivity (being overwhelmed with relevant and irrelevant stimuli) and hyposensitivity (ignoring both relevant and irrelevant stimuli) (see Eisenstein & Eisenstein, 2006; Eisenstein, Eisenstein, & Sarma, 2016, for more on homeostatic models of habituation and sensitization).

RNA Must Have Preceded DNA: The RNA World Hypothesis

The RNA world hypothesis states that all current life forms of life began with free-floating single strands of RNA. Although a single strand of RNA is relatively unstable (at least compared to double-stranded DNA), RNA is self-metabolizing and capable of replicating. RNA also contains four nucleotides: adenine (A), uracil (U), cytosine (C), and guanine (G) (note: in DNA, uracil is replaced by thymine [T],). Further, A bonds only with U, and C bonds only with G (note: in DNA, A bonds only with T). All directions for metabolism, replication, and other functions are coded within a complex (long) yet simple sequence of these four nucleotides, for example, CGATGCTACG, etc.

Ignored in the brief history of life presented in Chapter 1 is a more detailed discussion of the chemical precursors of life. Of course, this presents a new question of how and why these final four nucleotides that make up RNA formed in the first place. What scientific principles guided the formation of pre-nucleotic material, and why did these basic building blocks of life have affinities for one another? It seems that even the RNA world hypothesizers tend to ignore that there must have been a pre-RNA environment consisting of the chemicals necessary to form some basic nucleotides. Further, these chemical affinities and disaffinities must have been based basic atomic bonding principles. Thus, the original self-assembling molecules that formed the nucleotides creating the first strands of RNA were already constrained by fundamental chemical and subatomic physics principles. The study of these foundations of life from basic elements and chemicals is called *abiogenesis,* which itself is far from completely understood. Remember, however, at the outset of this chapter one definition of memory posits that it is a property of matter such that it maintains its state or structure over time. Certainly, this very broad definition applies to RNA and its nucleotides, and, thus, we might view the basics of memory as ultimately residing in the properties of subatomic particle physics.

Major Assumptions: Life's Basic Molecules Were Subject to Darwinian Natural Selection and Non-Associative Learning Principles

Although de Lorenzo focused on DNA, his selfish-metabolism hypothesis might arguably be true of RNA, nucleotides, and their chemical precursors. Further, it is likely that there were other complex molecules in the pre-RNA world, including nucleotides that failed to perpetuate because they did not expand their chemical landscapes or were not as successful at replication as other nucleotides that did successfully expand their landscapes. Molecular biologists regularly refer to the information and instructions coded within RNA or DNA as genetic memory, that is, memory that is passed down over generations. This coded information directs the synthesis of proteins providing cells with energy and guiding behaviors of the organism that ultimately leads to its successful replication. In this regard, the coded information is a form of memory that developed, presumably under natural selection. This memory form certainly meets the memory definition of Hertkorn and his colleagues (2013) for organic matter dissolved in oceans, and it also holds for RNA and its precursors. Also, experiences and conditions in parents can be passed on to their offspring yet lie outside traditional DNA transmission patterns. The latter phenomenon is included in a field called *epigenetics* (see Hedinger, 2016, for a review). However, few psychologists address the various evolutionary issues associated with genetic memories or the principles and conditions that lead to the acquisition of information coded in genes.

RNA and the Greek Goddess Athena

It is highly unlikely that RNA sprang out, fully grown, with just these four nucleotides, like Athena from the forehead of Zeus. I hypothesize that the original self-assembling molecules that formed pre-nucleotides and the final four nucleotides were subject to Darwinian natural selection. I also propose that the earliest molecular forms' ability to explore and exploit chemical landscapes in the interest of their own metabolism was guided (chemically and atomically) by at least two abilities: (1) to ignore stimuli not in their metabolic interest (habituation) and (2) to approach stimuli in their metabolic interest (sensitization).

Further, I purport that some chemical precursors to nucleotides and alternative nucleotide forms did not make the "final four" that came to define RNA, because they were not able to explore and exploit their chemical landscapes as well as the final four nucleotides. My opinion in this regard is influenced by complex systems scientist Stuart Kauffman (2000), who proposed that

> autonomous agents forever push their way into novelty—molecular, morphological, behavioral, organizational. I will formalize this push into novelty as the mathematical concept of an "adjacent possible," persistently explored in a universe that can never . . . have made all of the protein sequences even once, bacterial species even once, or legal systems even once. . . . Then the hoped-for fourth law of thermodynamics for such self-constructing systems will be that they tend to maximize their dimensionality, the number of types of events that can happen next. (p. 32)

Thus, from an evolutionary perspective, the central dogma of molecular biology (DNA codes RNA, RNA codes proteins) seems to overlook that (1) RNA must have preceded DNA, (2) the molecular precursors of nucleotides were based on chemical and atomic affinities and disaffinities, (3) there must have been other nucleotides before and during the composition of the final four, (4) the ability to explore and exploit new chemical landscapes (i.e., selfish metabolism) was the essential driving force in their successful evolution, and (5) this exploration of chemical landscapes was subjected to natural selection and conducted by the two types of non-associative learning, habituation and sensitization, which themselves were influenced by chemical-atomic prototypes of these two forms of learning. In summary, it may have been critical in a molecule's exploration of new chemical landscapes to ignore irrelevant stimuli (habituation) and to be sensitive to relevant stimuli (sensitization). Thus, I hypothesize that non-associative learning and its chemical and subatomic learning prototypes were the foundation for all subsequent learning and memory and, in all likelihood, the foundation for the formation of life.

Associative Learning

Associative learning is the establishment of a relationship (i.e., a memory) on the part of the organism, typically between some stimulus and a behavioral response. Associative learning consists of classical and operant conditioning. The earliest (in psychology) and most influential model of associative learning is the classical conditioning paradigm, attributed to the Russian physiologist Ivan Pavlov (1849–1936). Pavlov trained a dog to salivate (the conditioned response, or CR) to a sound like a bell (the conditioned stimulus, or CS) by pairing the bell with food (the unconditioned stimulus, or UCS). In the classical conditioning paradigm "natural" or "reflexive pairings," such as the sight and smell of food to a dog (UCS), elicit the unconditioned response (UCR), in the dog's case, saliva. It is important to note that the dog does not have to learn this association, as it is *said* to be a reflex. In the broader behaviorism paradigm, the organism's total behavior is called its *behavioral repertoire*. Behaviorists believe that an organism's behavioral repertoire expands on the basis of pairings of previous neutral or novel stimuli (i.e., stimuli that elicit no reliable or discernable response) with various kinds of UCS. In 1904, Pavlov won the Nobel Prize for Physiology/Medicine, primarily for his body of work on reflexes and digestion. As a side note, there is some debate over whether Pavlov actually used a bell as the CS, but it is known that he did use buzzers, metronomes, visual stimuli, and, likely, a bell. The other form of associative learning is operant conditioning, where an organism's behavior either increases or decreases in frequency based on its subsequent pairing with a stimulus. A reinforcing stimulus increases the probability of a response in the future. Punishment decreases the probability of a response. Operant conditioning probably reached its zenith in the latter part of the 20th century, owing to the influence of B. F. Skinner. His two most influential books were *Walden Two* (1948), where he presented a utopian society based on behavioristic principles, and *Beyond Freedom and Dignity* (1971), where he further advanced these ideas. Skinner's denial of the importance of neural mechanisms, brain functions, and internal mental thoughts ultimately limited the influence of the behavioristic paradigm, although many of his subsequent proponents did find it useful to address and integrate brain mechanisms and internal thoughts with operant conditioning techniques (e.g., cognitive behavioral therapy). However, he remained adamantly against such inroads into his theory, and publically called those who did so the worst imaginable name he could think of: *creationists!*

What I Never Asked and No Student Ever Asked Me!
One big elephant in the behavioristic-paradigm room that has been almost completely ignored is how a UCSUCR relationship is established in the first place. Because behaviorists generally downplayed or ignored internal states, brain mechanisms and networks, genetic influences, and genetic differences among individuals, they could *not* address how or why UCS–UCR pairings came to be established in the first place. According to the gold standard of all current knowledge, Wikipedia, a reflex is an involuntary movement in response to a stimulus. Additional explanations of reflexes are that they either innate or acquired, which explains nothing of their evolutionary origins. Long story made short: The UCS–UCR pairing was *learned*. And it may work like this: Imagine a variety of stimuli associated with food: primarily, their appearance (visual) and smell (olfactory). Even prelife forms were exploring their chemical landscapes in search of compounds that enhanced their metabolism, which increased their probability of reproducing. Errors in DNA replication are mutations (e.g., faulty enzymatic reactions result in DNA nucelotide changes), which often lead to phenotypic (either structural or behavioral) variations. Note that phenotypes are a result of individuals' genes interacting with their environment. In the case of responding to food stimuli, it would be the phenotype of behavioral response time. Now, imagine a variety of response times based on variations inherent in organisms' DNA. Those that are faster to respond to stimuli which herald food or chemicals that enhance their metabolism (and subsequently their reproductive success) are more likely to sustain themselves and pass those genetic predilections to subsequent offspring. Over generations, the natural selection for the genotype of a faster response time will increase in the population and become fixed in DNA (i.e., genetic memory). The latter process is also called the *Baldwin effect* (i.e., adaptive learning that increases fitness).

The Baldwin Effect

In 1896, psychologist J. Mark Baldwin (1861–1934) proposed a "new factor" in evolution. It was the notion that when organisms encounter new environments or when extant environments change, those organisms that can learn to adjust more quickly will be favored by natural selection over those that are less flexible or slower. Over many generations, the successful offsprings' continued and sustained behavior would ultimately change or shape the evolution of that species. Baldwin's new factor was this interaction between an organism's environment and its genetic code, and it became known as the Baldwin effect. Presently, it is accepted as a theoretically viable mechanism in evolutionary theory. In the present context, it helps to explain why all reflexes are learned and why the "innate versus acquired reflex" dichotomy is ultimately specious, as *all* reflexes are learned initially.

Another example of a reflex is from an observation made by Charles Darwin (1872) when visiting the London Zoological Gardens. He said that he pressed his face up against the glass cage of a puff adder with the full intent of not recoiling when the adder struck the glass. However, despite his full resolution not to move, he found himself jumping "a yard or two backwards with astonishing rapidity" (p. 38). Further, he wrote, "My will and reason were powerless against the imagination of a danger which had never been experienced" (p. 38). Darwin correctly surmised that his reaction was an ancient reflex. The neural circuitry of the reflex is now well known. The visual image of the snake reaches the amygdala, which sends out a direct signal to motor neurons to instigate the recoil reflex, and this process initially bypasses the cerebral cortex. Of course, this behavior was not encoded directly in an ancestor in one trial. However, it must have been beneficial for most ancestral organisms to avoid snakes quickly. Ancestral organisms that were able to avoid a striking snake more quickly than others were ultimately at a reproductive advantage. The genes that caused even a slight tendency to recoil faster than peers were passed on to their offspring. Over a very large number of generations and perhaps millions of years, we, like Darwin, can observe reflexes (or genetic memories), and only the process of natural selection is required to understand their acquisition. What DNA mutations provided in this context was a variation in response speed. Ultimately, what natural selection offered was a bias for the genetic codes that favored the speedy—that is, the Baldwin effect. Voilà, over great spans of time, the result is a UCS–UCR pairing, which is so often neglected in explications of the classical conditioning paradigm.

The Evolution of Associative Learning

As noted in Chapter 1, eukaryotes (cells with a nucleus) appeared about 3.4 to 2 billion years ago (the timing of their arrival is debatable), and their more complicated structure and metabolism likely required more complex genetic machinery. Thus, the double-stranded more stable DNA molecule appeared, which carried these instructions for the coding of RNA, which in turn coded proteins. These proteins then coded the metabolic and other information for these cells to survive and reproduce. About 1 billion years ago, eukaryotes began to assemble into multicellular forms. Why did natural selection begin to favor multicellular groups like comb jellies and sponges, which date from about 600 million years ago in the fossil record? It is a difficult question to answer, although the answer might reside in a resulting structure that is more efficient than individual cells in food acquisition and food consumption. This structure probably aided reproduction, and various parts (top, bottom, side, middle) could begin to specialize, heralding the beginnings of the animal kingdom. Unicellular eukaryotes are classified as protozoans. Comb jellies and sponges (multicellular eukaryotes) are considered the basic group (clade) of metazoans (i.e., animal life).

The Cambrian Explosion of Animal Life

Within a relatively short period of evolutionary time after the arrival of multicellular eukaryotes, a major event occurred in the number, diversity, and complexity of animal life. It is called the *Cambrian explosion,* which began about 545 to 520 million years ago. This amazing array of fossils suddenly appears in phylogenetic history. Some of the finest specimens first found were in Wales, UK (the Latin names for Wales is *Cambria,* hence the name of the explosion). A confluence of multiple factors may have accounted for this sudden expansion of animal life, including biological, climatological, tectonic, geochemical, and others. One suspected important factor is an increase in oxygen about 635 million years ago, perhaps in part from the abundance of decaying prokaryotes and eukaryotes. During the Cambrian period, flatworms appeared with bilateral symmetry (they are called *bilaterians*), rudimentary nervous systems, "brains," photosensitive cells ("eyes"), mouths and anuses (same organ), and the ability to swim about in their environment. Thus, from just the time of the appearance of comb jellies and sponges, eukaryotic cells had already specialized into (1) sensory cells, (2) motor cells, and (3) cells to interpret sensory information for appropriate guidance of motor movements, that is, brains or neurons (cells of the brain). It may be argued that brain cells primarily evolved for effective movement, based on the interpretation of information from sensory cells. By moving, these organisms could approach and sense appropriate things to eat, move away from or avoid toxins and obstacles, avoid being eaten, and approach (or avoid) others in order to reproduce (or not). This latter ability resulted in a more successful (selective) means of reproduction. These brain cells had already developed on the leading end of the organism, no doubt to better guide movements. Interestingly, modern executive functions (goal-directed behavior, decision-making, planning, organizing, strategizing, inhibiting, etc.) are also located in the anterior portions of the brain (frontal lobes). (It is interesting to note that Athena sprang from the *forehead* of Zeus, although it may not have been mere coincidence. Many of the ancient Greeks, including Socrates, Plato, and Hippocrates, recognized the brain as the "seat" of thinking, although not necessarily the frontal lobes. However, that Athena is viewed as the goddess of wisdom, rationality, and intelligence *is* an interesting coincidence.)

Currently, scientists study flatworms, such as planaria and amphioxus, as possible representatives of the Cambrian simple flatworms. Because planaria and amphioxus demonstrate associative learning, both classical and operant conditioning, it is strongly suspected that Cambrian metazoans were capable of these same associative learning mechanisms. Further, it is suspected that unlimited associative learning may have been one of the chief causes of the Cambrian metazoan explosion (e.g., Bronfman, Ginsburg, & Jablonka, 2016; Ginsburg & Jablonka, 2010). It is said to be "unlimited" because an organism could learn more than the span of its life would allow.

Did Associative Learning in the Cambrian Cause a Speciatic War?

Given that associative learning helps select for more complex organisms, natural selection favored organisms that had enhanced learning abilities (as previously noted with the Baldwin effect), more elaborate central nervous systems, enhanced adaptability, and greater reproductive success. What appears to have happened in the Cambrian explosion, particularly because of the bilaterians' movements, was that their movements caused a highly novel new natural selection scenario: There was a new group of predators and a new group of prey. Trying to eat and trying to avoid being eaten appears to have been selected for greater discriminatory power of the senses, additional senses (olfaction, visual, haptic [touch, including electrical sensations], audition, and gustatory), and more sophisticated motor appendages. Further, morphological variations were naturally selected, which led to better predation but also better defenses against predation, for example, harder body parts, spines, toxins, scales, teeth, horns, spikes, camouflage, flight, burrowing, and diurnal or nocturnal living. The variability in defenses implies that organisms could also become more effective survivors and reproducers by finding well-suited environmental niches, such as living in trees, water (oceans, lakes, rivers, hot deep-ocean vents, etc.), deserts, mountains, plains, and savannahs. Note that it would not behoove all organisms to select the same niche, but it would aid some percentage of the general population of organisms to find their appropriate environmental niche.

The Relationship between Associative and Non-Associative Learning: The Exaptation of Associative Learning from Non-Associative Learning

As noted in Chapter 1, an *adaptation*, in Darwinian terms, is the development of a trait (morphological or behavioral) that has been naturally selected because it affords greater fitness, making it more likely that an organism will survive and reproduce. To follow a classic example, feathers were originally an adaptation to the environment to maintain body temperature (thermoregulation). As noted previously, a physical or a behavioral trait is referred to as an organisms' *phenotype* (and its genetic makeup is its *genotype*). An *exaptation* refers to the change in phenotypic function of a pre-existing structure or feature from its original adaptive value with a new function. Thus, feathers, in addition to their adaptive value for thermoregulation, became *exapted* for flying. At this point, I will invoke Occam's razor (make no more assumptions than necessary to account for a phenomenon). I hypothesize that the molecular (chemical and atomic) mechanisms that underlie habituation and sensitization (which may have operated in at least proto-forms since the origins of earth or even earlier) were *exapted* to solve this problem of instantiating or preserving into memory something that was learned. In non-associative learning, a relationship is established between a repeated stimulus and a response (it is bit of a misnomer to label it non-associative). The same is true of the two forms of associative learning, classical and operant conditioning. At their core, both non-associative and associative learning are establishing stable, recallable relationships between stimuli and responses. At a neural level, changes in the protein composition of individual neurons or changes in neural networks (how neurons are "wired" together) improve the communication between sensory neurons and motor neurons, which allows more intelligent interactions with an organism's environment. Thus, organisms possess the ability to memorize, which allows the use of past experiences to anticipate future events and rewards, and the ability to discriminate between stimuli and classes of stimuli. Further, scientists consider associative learning the basis of all animal cognition and even the beginnings of consciousness.

Preliminary Conclusions

1. Modern human memory systems are based on non-associative and associative learning principles.
2. Non-associative learning principles (habituation and sensitization) are based on more basic chemical and subatomic particle affinities and disaffinities.
3. The origins of life can be traced to a pre-nucleotidic world, where selfish metabolism (the ability to exploit new chemical landscapes) and reproduction were paramount.
4. The pre-nucleotidic world operated on the basis of non-associative learning principles, which successively guided the formation of varieties of more complex nucleotides. Through natural selection (the ability to exploit beneficial chemical landscapes, which enhanced an organism's metabolism and enhanced its reproductive success), only four nucleotides survived to form the basis of RNA in prokaryotes, which later became modified to form double-stranded DNA eukaryotes.
5. The Cambrian explosion of metazoan, beginning about 545 million years ago, resulted from the development of virtually unlimited associative learning (classical and operant conditioning).
6. Associative learning developed on the basis of the exaptation of non-associative learning principles, that is, the establishment of an association between a stimulus and a response.

Baddeley's Working Memory Model

One predominant memory model that subsumes nearly all memory categorizations is *working memory*. There is, however, some confusion in the cognitive literature about the concept of working memory. The generic definition of working memory is that it is what one can actively hold in one's mind despite interference. The present discussion will focus on a second, more specific definition of working memory but much broader in scope, and it is the memory model elaborated on by experimental psychologist Alan Baddeley and his postdoctoral fellow Graham Hitch (Baddeley & Hitch, 1974; see also Baddeley, 2001, 2012). Baddeley's model has received by far the most attention and empirical support in the cognitive literature over the past five decades. See Fig. 2.1 for my modification of his model (Baddeley originally used boxes).

By the mid-20th century, psychologists had come to a general agreement that there were two kinds of memory, short-term and long-term. Back then, short-term memory was said to be material acquired recently, as in the previous 2 minutes, and then recalled. This research was almost exclusively devoted to the study of an acoustic, limited-capacity verbal store for words or

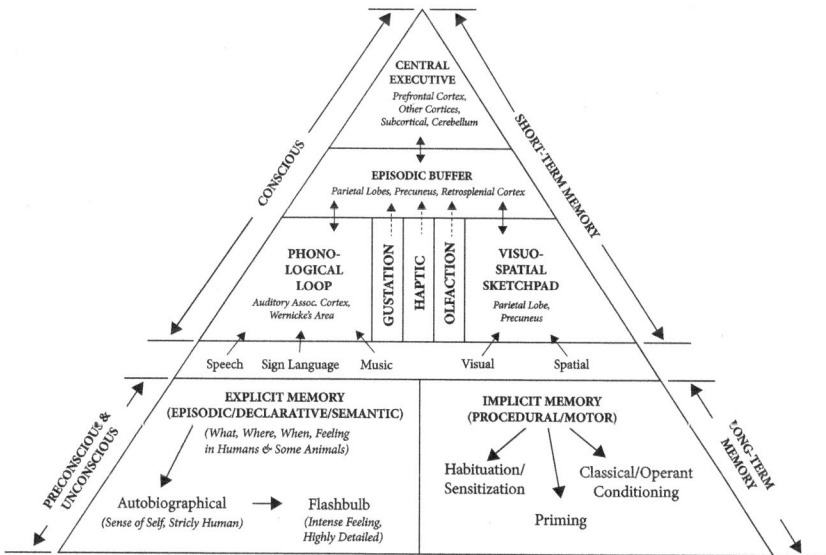

Fig. 2.1 Baddeley's working memory model.
Source: James M. Hicks.

numbers. Long-term memory was information that had been learned, but the recall was delayed by hours, days, or even longer. This research also was mostly devoted to acoustic verbal material. By the early 1970s, psychologists like Baddeley had recognized the limitations in this two-component model of memory and its dominant focus on verbal information.

Baddeley and Hitch's initial work then was to develop a more comprehensive cognitive theory that accounted not only for the standard operations of short-term memory but also for how and where memory is established and how it is relates to long-term storage. As noted earlier, this comprehensive working memory model (Baddeley & Hitch, 1974) had a profound effect on the field of brain and memory research. Their initial model included an attentional, panmodal controller or central executive, and two subsystems, the phonological loop and the visuospatial sketchpad. Baddeley (2000) expanded the central executive's functions by adding a new component, the episodic buffer, which integrates and temporarily stores information from the other two subsystems. The phonological loop contains two elements, a short-term phonological store of speech and other sounds, and an articulatory loop that maintains and rehearses information either vocally or subvocally. Baddeley thought one of its important purposes was language acquisition and comprehension. The visuospatial store was hypothesized to involve the maintenance and integration of visual ("what" information, like objects) and spatial ("where" information, like location in space) elements and a means of refreshing it by rehearsal.

The Central Executive and Long-Term Memory

Baddeley (2012) currently views the *central executive* as a system of varying functions including attention, active-inhibition, decision-making, planning, sequencing, temporal tagging, and the updating, maintenance, and integration of information from the two subsystems. Some brain function models present working memory as simply one subcomponent of the various functions of the prefrontal cortex and as part of a network that involves all other lobes of the brain, many subcortical structures, and the cerebellum. There is also a raft of new evidence from empirical studies that intimately (if not synonymously) links Baddeley's central executive component to the traditionally defined aspects of *executive functions* from the clinical neuropsychology literature (for a review of contemporary working memory models and empirical evidence, see

Baddeley, 2012; Osaka, Logie, & D'Esposito, 2007). This clinical literature primarily studies patients with varying forms of brain damage and brain dysfunction in order to understand how the brain works. Thus, executive functions are approached from the perspective that if the prefrontal cortex is damaged by head injury, stroke, or other means, then such patients will show attention problems, disinhibition, indecisiveness, decision-making difficulties, poor planning, and an inability to see how things go in their proper order.

Important in the present discussion is that in most conceptions of the central executive or executive functions, working memory not only serves to focus attention and all of the aforementioned other functions but also serves as the chief liaison to long-term memory systems, including language comprehension and production. Further, the generic definition presented at the outset of this discussion is still appropriate to Baddeley's multicomponent working memory model—that is, the amount of information one can hold in one's mind despite interference. In empirical research of Baddeley's model, this amount of information is referred to as *working memory capacity*. Research has also shown that working memory capacity is moderately or strongly related (i.e., predictive) to a wide variety of higher cognitive measures, including reading and listening ability, language comprehension, vocabulary, reasoning, writing, computer-language learning, and many others (e.g., Engle & Kane, 2003). Importantly, working memory capacity is strongly correlated with intelligence, both crystalized intelligence (learned information) and fluid intelligence (novel problem-solving; Kane, Hambrick, & Conway, 2005). Working memory capacity, whether measuring individual components like executive functions, phonological storage capacity, or visuospatial sketchpad capacity or measuring a composite of the components, is also highly heritable and polygenic (i.e., involving multiple genes), as demonstrated by empirical research (see Wynn & Coolidge, 2010, for a review; also Friedman et al., 2008). I will return to the importance of the heritability of working memory in later chapters. Before I survey yet another categorization of long-term memory, it is important to review Baddeley's other three components of working memory.

Phonological Loop and Its Evolution

The *phonological loop* is intimately involved in language use. Baddeley hypothesized that the phonological loop has two components, a brief, sound-based storage that fades within a few seconds, and an articulatory processor. The latter maintains material in the phonological store by vocal or subvocal rehearsal. Spoken information has automatic and obligatory access to phonological storage, and Baddeley therefore hypothesized that the phonological store evolved principally for the demands and acquisition of language. Baddeley and Logie (1999) wrote that "the phonological loop might reasonably be considered to form a major bottleneck in the process of spoken language comprehension" (p. 41).

Sounds that are held in the phonological store can be vocally repeated or subvocally repeated by means of the articulatory processor. This repetition will relegate those sounds into long-term memory storage if they are rehearsed enough, the person is adequately motivated, and the sounds have a sufficient emotional salience. A strong motivation to memorize, or an elevated emotional meaning (e.g., someone to whom you are attracted has an unusual first name), will increase the likelihood that sound will be successfully transferred into long-term memory. Sounds can be relegated to long-term memory even if they have no initial meaning. For example, repeating "won-due, era-due, muru" over and over, either vocally or subvocally, will eventually transfer them to long-term memory without any meaning attached to them whatsoever (they are the phonetic sounds for the numbers 1, 2, 3 in a southern India Dravidian language, Kannada). Baddeley, Gathercole, and Papagno (1998) have noted the important role that the phonological loop plays in the acquisition of a child's native language and in adult second-language learning. The process of vocal and subvocal articulation also appears to play a critical role in memorizing visual stimuli (Baddeley's visuospatial sketchpad), for example, thinking or saying, "Ah, a small blue chair!" The functions of the phonological loop also help explain why brain-damaged patients who have lost their ability to repeat sounds vocally can still memorize them. However, those patients who cannot create a sound or speech through the phonological loop cannot memorize new material (Caplan & Waters, 1995).

On the Influence of Mammal and Primate Brains on Modern Human Brains

As noted in Chapter 1 and will be noted in Chapter 3, modern human brains have been strongly shaped by virtually all animal forms throughout evolution, but they were especially impacted by changes in mammalian brains about 200 million years ago and then by changes in primate brains beginning about 65 million years ago. Aboitiz and his colleagues emphasized that an evolutionary enhancement in phonological storage's carrying capacity was likely derived in primates because of their highly vocally mediated social activities (Aboitiz, 2017; Aboitiz, Aboitiz, & Garcia, 2010; Aboitiz, Garcia, Bosman, & Brunetti, 2006). Further, they postulated that language has evolved primarily through the expansion of this same increase in phonological storage capacity, "which has allowed the processing of sounds, conveying elaborate meanings and eventually participating in syntactic processes" (Aboitiz et al., 2006, p. 41). They believe that this expanding memory system allowed for more complex memories, representing multiple items to be manipulated combinatorially. However, they thought these manipulations demanded significant working memory resources (demands on functions of the central executive). Interestingly, the combinatoriality of sounds is virtually unique to humans, with rare exceptions. Putty-nosed monkeys have at least two distinct vocal predators alarm calls, which, when taught to their young, elicit two different kinds of evasive actions. However, when those two sounds are combined, a group behavior is elicited, which seems to mean something on the order of "Let's move to another site for food" (Schlenker, Chemla, Arnold, & Zuberbühler, 2016; Scott-Phillips & Blythe, 2013). It appears that one critical prerequisite of combinatorial communication is not only an organism's sociocognitive capacities but also aurally directed communication, that is, intentionally outward expression of sounds toward others and recognition by others that these sounds represent intentions.

Visuospatial Sketchpad and Its Evolution

The *visuospatial sketchpad* is a temporary store for the maintenance and manipulation of visual and spatial information. It plays an important role in spatial orientation and solving visuospatial problems. Studies with brain-damaged patients and with healthy adults appear to demonstrate that the visual and spatial processes comprise separate memory systems, although the visuospatial sketchpad is assumed to form an interface between the two systems (Shah & Miyake, 2005). Visual information can also be integrated with other senses, and the most common primary sense that visual information is integrated with appears to be auditory. However, the number of empirical studies of the visuospatial sketchpad have lagged behind studies of the phonological loop. Because extant flatworms have rudimentary photosensitive "eyes," it may be surmised that the visual aspect of the Baddeley's visuospatial sketchpad evolved well before the phonological component. Whether the spatial component co-evolved with the visual component is debatable, although based on prokaryotes like bacteria, the spatial component may have evolved subsequently.

Episodic Buffer

Baddeley (2001) came to recognize that he had failed to attribute the central executive with its own storage capacity, emphasizing instead its attentional nature. So he proposed an *episodic buffer* as the temporary storage component of the central executive. He endowed the episodic buffer with the ability to bind together and integrate the two subsystems, the phonological loop and the visuospatial sketchpad, as well as traces from long-term memory via a multimodal code. By attending to multiple sources of information simultaneously, the central executive is able to create models of the environment that themselves can be manipulated to solve problems and even plan future behaviors and alternative strategies, so that if a plan fails, another may be chosen or generated. The evolutionary value of the episodic buffer will be discussed shortly.

A Long-Term Memory Dichotomy: Explicit Memory vs. Implicit Memory

It has been said there are only two kinds of people: those who categorize people and those who don't (Benchley, 1920). Cognitive psychologists appear to fall into the first category or are tyrannized by it. So, as previously noted, long-term memory is divided into explicit memory (declarative or semantic [meaning]) and implicit (procedural) memory. The *explicit memory* category has the connotation of being obvious, clearly stated, or associated with facts, words, meanings, and events.

One common subcategory of explicit memory is *episodic memory,* which is a memory for a visual, coherent, story-like reminiscence for an event that contains information about where and when the scene took place, what things were in the scene, and some emotional feeling about it (its emotional valence). Most cognitive scientists grant episodic memories to animals like birds (scrub jays) and foxes (e.g., Allen and Fortin, 2013), so episodic memory is not unique to humans. One subcategory of episodic memory, *autobiographical memory,* is restricted to humans, because it contains a sense of self in the memory, as it is unlikely (but not impossible) that a bird or a fox has a sense of self in their recall (see the end of Chapter 10 for a philosophical discussion on an animal's sense of self). An additional subtype of autobiographical memory is a *flashbulb memory,* which is usually associated with a very strong emotional valence. Often they consist of highly traumatic events, but they can also consist of happy events. It is interesting to note that each generation (cohort) will often have a common flashbulb memory. American baby boomers (born between 1946 and 1964), for instance, will ask one another where they were when they heard that President Kennedy had been assassinated (Nov. 22, 1963). For Generation X'ers (1965–1981), it might be the events of September 11, 2001. These flashbulb (the very name dates itself) memories are, of course, not restricted to people in the United States. When I was teaching graduate students at the Indian Institute of Technology Gandhinagar in 2016, their common flashbulb memory was the communal riots in 2002 in the city of Ahmedabad in western India, where many residents died or were injured.

Implicit Long-Term Memory

Implicit memory is a memory for a procedure, skill, or knowledge that is acquired through practice or repetition. It is also commonly referred to as procedural memory. These types of memory are often (but not exclusively) motor actions, like learning to ride a bicycle, learning how to make stone tools (stone-knapping), or learning to play the piano. Unlike declarative memories, the procedural or motoric kinds of memories may take hundreds or thousands of trials or years for a person to become proficient. If I asked you to memorize the word "won-due" (that's the phonetic spelling of the Kannada word for "one"), you could probably do it in one or two repetitions. If I wanted to ensure that you'd remember it in a month, I could make you repeat it 20 times or so. Another possibility for recalling the word in a month might be to increase the emotional valence of the word and situation. I could do so by offering you $1,000 if you get it correct later.

However, procedural memories are learned very differently. I once saw an 18-year-old guitar player in the hills of Georgia play perfectly the song *Rocky Mountain Way*, by Joe Walsh. It seemed to me he played it note for note, sound for sound, including the sounds Walsh's guitar frets made when his fingers moved across them. As it was a small venue, I went up afterward to congratulate him, and he was hardly thankful. He was almost hostile, as he told me he had done virtually nothing else for 6 months except sit in his trailer trying to master that song. Interestingly, behaviorists have a label for his reaction, *post-reinforcement pause*, where the energy expended over many trials required to master a reward (in this case to master a guitar skill) is subjectively viewed as inordinately more than the value of the reward (or the skill). Also, in contrast to the nature of explicit memories, even if I increased the emotional valence of the task, you could not learn to play that song without thousands and thousands of repetitions. One final thought: The declarative–procedural memory distinction has a neural basis, as both forms use relatively independent neural pathways; selective types of brain damage may affect one type of memory but not the other. To become a permanent long-term memory, declarative memories require functions of the hippocampus (as will be explained in greater detail in Chapter 8). To be made permanent, procedural memories do not require the hippocampus.

Explicit Long-Term Memory: Episodic Memory and Autobiographical Memory Revisited

Memories of a person's past have long been examined and studied by psychologists. William James (1890), in his classic book *Principles of Psychology*, discussed such "recollections" in great detail, as did educational psychologist Edward Thorndike in 1905. Many attribute the specific term *episodic memory* and more modern research on it to experimental psychologist Endel Tulving (2002), whose work on episodic memory began in the 1960s. He established a distinction between knowing something, like a fact, detail, or meaning, and remembering something, like a recollection from one's past. The former has come to be referred to as *semantic memory* but also is often distinguished as a subcategory of declarative or explicit memory. Episodic memory is also a subcategory of declarative or explicit memory but has characteristics that make it distinctly different. For example, a memory for past experiences is usually visual, and if sufficiently meaningful or emotionally arousing, it can persist for a very long time (like recalling traumatic or joyful childhood experiences). These memories are usually recalled as short experiences (much less than a minute) and are re-experienced subjectively, usually with an emotional or affective attachment (which can be positive, negative, or neutral, but a neutral response means that the experience will be more vulnerable to being forgotten).

Cognitive scientist Roger Shepard (1997) noted that one of the critical evolutionary advantages of recollections was "thought experiments." He stated that every real-world experiment to solve some problem had to have been preceded by corresponding thought experiments. However, he envisioned that they drew on "countless" prior real experiences, which had been recombined in one's mind to produce a new solution. The advantage of this internalized knowledge is that one could avoid "trial and possibly fatal error" (p. 24). Although Shepard did not specifically mention episodic memory, it is evident that he appreciated one of its chief evolutionary advantages.

Tulving (1972) proposed the term *autonoesis*, which is unique to humans and refers to the ability to form a special kind of consciousness in which one becomes aware of the subjective nature of time. It is this form of consciousness that allows humans to travel mentally back and forth in time—that is, recalling and even manipulating past events and simulating alternative future events. Mental time travel, by way of episodic processes, helps a person become aware of the subjectivity of time. Tulving (2002) notes,

> This awareness allows autonoetic creatures to reflect on, worry about, and make plans for their own and their progeny's future in a way that those without this capability possibly could not. *Homo sapiens*, taking full advantage of its awareness of its continued existence in time, has transformed the natural world into one of culture and civilization that our distant ancestors, let alone members of other species, possibly could not imagine. (p. 20)

Interestingly, Baddeley (2001) views his own proposed episodic buffer in working memory as intimately tied to Tulving's concept of episodic memory, although Baddeley views the episodic buffer as primarily a temporary and integration storage buffer, whereas he sees Tulving's episodic memory as primarily a system concerned with long-term storage. Tulving also placed an emphasis on the recent evolution of episodic memory, while acknowledging that many animals are capable of remembering the "what, where, and when" of an episode. In my opinion, the distinction between human episodic memory and animal episodic memory (e.g., Allen & Fortin, 2013) is a valid one, as Tulving places a proper emphasis on the notion of one's "self" in an episodic memory, which I highly doubt scrub jays do. As Tulving noted, some animals have a temporary notion of their own "existence," as evidenced by animals recognizing themselves in a mirror, but that existence is confined to the ephemeral present and is unlikely to be carried forward as a stable sense of self. Also, animals are in all likelihood not capable of becoming subjectively aware of the ability of traveling backward or forward through time.

On the Critical Evolutionary Importance of Episodic Memory

It is obvious that human memory is far from perfect. Cognitive psychologist Daniel Schacter (2012) and his colleagues (e.g., Schacter & Addis, 2007; Schacter, Guerin, & St. Jacques, 2011) have viewed the imperfections of memory (e.g., forgetting, misremembering, biases, false memories, confabulation [mixing fact and fiction inextricably], etc.) not as evidence of a defective or flawed system but as evidence of a "dynamic memory system that flexibly incorporates relevant new information" (Schacter, 2012, p. 2). Further, Schacter and his colleagues propose that the original function of episodic memory was *not* the recall of an episode for the sake of recalling that episode, and it was certainly *not* for the purpose of recalling that episode *exactly* as it happened. They hypothesize that episodic memory was evolutionarily adaptive because relevant episodes could be recalled, reconstructed and reassembled, and then used to simulate alternative future scenarios and outcomes in those various scenarios. As Shepard (1997) had earlier noted, these simulations (i.e., thought experiments) were much safer than actual trial and sometimes fatal error.

Baddeley (2012) also proposed that greater working memory capacity, particularly with regard to episodic memories, would allow for the formulation of mental models that were more likely to be successful as future behaviors. Fascinatingly, Baddeley (2012) even proposed the ability to imagine an ice hockey–playing elephant (indeed, after reading that, it is hard not to imagine it) and whether the elephant would be better as a goalie or as a body-checking defensive player! It is reasonable to assume that greater working memory capacity would allow for the reflection and comparison of multiple past experiences. This might allow an individual to actively choose a future action or create an alternative action, rather than simply choosing the highest path of probable success, as Baddeley has noted. Although an individual would still be better off (compared to one without benefit of past experience) choosing alternatives simply based on the past successes and failures (an example of an inflexible anticipatory process), an individual who could actively compare multiple simulated scenarios and combine them into a novel action would undoubtedly have a strong advantage. The latter ability to solve novel problems with novel solutions is again the essence of fluid intelligence, and studies that demonstrate the strong association between working memory capacity and fluid intelligence are therefore not surprising.

Back to the Australopithecines, Habilines, and *Homo erectus*

If basic episodic memory is granted to modern animals, then Lucy and other australopithecines about 3 million years ago and the habilines (*Homo habilis* and others) about 2.5 million years ago must have had episodic memories as well. However, the australopithecines and habilines largely maintained an ape-like way of arboreal life with some bipedalism but still arboreal locomotion. Thus, I suspect their reliance on episodic memories was largely unconsciously invoked, highly ephemeral, and contained little or no sense of self. In other words, their episodic memories were probably not similar to the strong sense of self in modern human autobiographical memories. They probably did not sit back and consciously recall particularly satisfying episodic memories in which they saw themselves actively within the memory. Further, it seems highly unlikely they possessed autonoetic awareness. I do not think that when they recalled these episodes they understood that time was subjective and relative and that one could travel back and forth in time and could even imagine future scenarios. I would imagine that they were incapable of Shepard's thought experiments. I suppose they could conduct live experiments with various ways of completing some task, but undoubtedly they could not conduct these experiments in their minds prior to the actual experience.

Remember, as noted in Chapter 1, *Homo erectus* appeared about 2 million years ago, and their appearance represented a major grade shift in hominin evolution. *Homo erectus* had left the trees and adopted a completely terrestrial way of life. They were fully bipedal and no longer mostly ape-like in their behavior nor in their size. They were not yet fully human, but their brains had increased substantially to nearly the lower limits of modern brain sizes. They were as tall or taller than modern humans but perhaps leaner, owing to their active lifestyle. Their stone tools were also dramatically different, and their toolkit not only included sharp flakes and pounding stones but also the symmetrical, bifacial handaxe, a design that would persist for the next million years or more. My colleague, archaeologist Thomas Wynn, and I have written extensively about the cognitive leap that was characteristic of *Homo erectus* (Coolidge & Wynn, 2005, 2018; Wynn & Coolidge, 2007a,b, 2010), so I will not belabor that point here.

Most anthropologists and some linguists do grant that *Homo erectus* may have had some form of language. If modern vervet monkeys have different sounds for different predators (Price et al., 2015) that elicit distinctively different responses, *and they teach these sounds to their young*, then *Homo erectus* may have had a vocabulary more extensive than that of extant apes and monkeys, especially with their much larger brains and undoubtedly much larger group sizes. Thus, I would hypothesize that *Homo erectus* was capable of consciously recalling previous episodic memories relevant to a present task. As their territorial range expanded over 10 times the range of the australopithecines, I would imagine the recall of scenes along their explorations would be highly useful and adaptive. I can also imagine that they may have been able to recall and describe these episodes to their group's members, although that may be too speculative. However, I cannot imagine that they would recall these episodes *without* any sense of self in their recall. I think it much more likely that *Homo erectus* became the first hominid to have some autobiographical representation of self in their episodic memories. Would they have had an autonoetic sense—that is, time is relative and memories may become a path to travel along? Perhaps not. Could they have consciously assembled and constructed future scenarios and alternative ways of doing things before they actually did them? Maybe, or maybe not. However, I do suspect that when they did consciously make a decision to act or to solve a novel problem or while wayfaring, they were drawing on their episodic memories with some stable sense of self. This recall and reconstruction required extensive brain networks and involved various regions of the brain, which will be discussed in greater detail in Chapter 3.

Summary

1. Modern human memory systems are based on non-associative (habituation and sensitization) and associative (classical and operant conditioning) learning principles.
2. Non-associative learning evolved first, guiding the formation of increasingly more complex molecules. The principles underlying habituation and sensitization are based on basic chemical and subatomic particle affinities and disaffinities.
3. In a prelife and pre-nucleotidic world, selfish metabolism and reproduction were paramount. Through natural selection, only four nucleotides survived to form the basis of RNA in prokaryotes, which later became modified to form double-stranded DNA eukaryotes.
4. The Cambrian explosion vastly expanded the animal kingdom, beginning about 545 million years ago. It resulted from the development of virtually unlimited associative learning. Associative learning developed on the basis of the exaptation of non-associative learning principles, that is, the establishment of an association between a stimulus and a response.
5. Baddeley's multicomponent working memory model aids in the understanding of the evolution of cognition.

3

An Introduction to the Brain

Life, Brains, and Consciousness

What is the definition of life? In Chapter 2, the point was made that coagulating molecules required selfish metabolism and a means of replication to evolve. Both of these characteristics may be considered essential and absolute, in that life cannot continue to exist and evolve without them. In that regard, an inchoate foundation is established for a general definition of life. Hungarian theoretical biologist and chemist Tibor Gánti (1975) proposed an abstract model of the minimal requirements of life, which he called a *chemoton*. He noted that living systems differ from simple mixes of organic molecular compounds in the former's "special organization manner" (p. 15). He hypothesized that chemotons and thus living systems began with the formation of cells (i.e., prokaryotes). Cells, he stated, were chemical self-organizing "supersystems" composed of three subsystems: cytoplasm, genetic material, and a cell membrane. His minimal living-system requirements for a cell and its subsystems (with minor modifications by this author) consisted of (1) an inherent unity and stability (aided by the cell's membrane); (2) self-metabolism (capable of synthesizing source compounds, as needed); (3) an information, memory, and guidance system (aided by the ability to acquire information, store it in a genetic code, and impose it); (4) growth, replication, and death; and (5) evolution by a hereditary system (the genetic material), which allows for mutations and subsequent phenotypic diversification.

As noted in Chapter 2, a prototype for brains began with the nucleus of eukaryotes, and brains became the primary organ for interpreting sensations about a cell's environment and conducting appropriate movements based on those sensations. However, I suspect no one would attribute *consciousness* to even the initial eukaryotes, such as comb jellies and sponges, which formed the basis for all animal life at about 600 million years ago. Furthermore, what is an acceptable definition of consciousness?

Just as Gánti had hypothesized the requirements for a transition from organic molecules to life, Bronfman, Ginsburg, and Jablonka (2016) proposed

Evolutionary Neuropsychology. Frederick L. Coolidge, Oxford University Press (2020).
© Oxford University Press.
DOI: 10.1093/oso/9780190940942.001.0001

a transition marker from life to consciousness. Their definition for minimal consciousness (also known as *basic consciousness* or *first-order consciousness*) was "simple subjective experience . . . experiencing of a feeling of comfort or fear, or . . . colour" (p. 9). Their transition marker was the ability of organisms to learn associatively (classical and operant conditioning) during the Cambrian explosion of life (Ginsburg & Jablonka, 2010, 2015). They noted that non-associative learning (habituation and sensitization) was essentially limited to a small number of actual associations between stimuli and responses in the life of an organism (as I noted earlier, why the term *non-associative* when a reliable response is established between a stimulus and a response?). They proposed that associative learning, in contrast, was essentially *unlimited*, and thus they coined the term *unlimited associative learning*, where the number of associations that can be learned and recalled within and between sense modalities exceeds not only the lifespan of an organism but also by far the number of individuals in a given population. Further, although the biological nature of an organism constrains associative learning to some extent, "the number of possible learned associations is vast, and learning-based plasticity is never fully exhausted" (Ginsburg & Jablonka, 2015; p. 59). Thus, unlimited associative learning was Bronfman, Ginsburg, and Jablonka's transition marker to conscious brains. Interestingly, in this regard, consciousness may be considered an exaptation of the basic learning adaptations, which helped to form living things.

Recognition That a Brain Controls Life Functions

When did hominins first realize that a brain controlled life functions? In the 1920s, anthropologist Raymond Dart (1949; C. K. Brain, 1972) recognized that some australopithecine skulls, as well as baboon skulls, had been bludgeoned by bone weapons, undoubtedly by other australopithecines. This finding and claim may serve as an indication that the australopithecines, perhaps as far back as 3 million years ago, had some recognition that brains controlled life functions. In a major irony, the earliest written evidence of investigations into brain–behavior relationships comes from Egyptian hieroglyphics dating to about 3,700 years ago or even earlier. Although Egyptian physicians were superstitious, believing that the favor or disfavor of the gods and the influence of evil spirits resulted in diseases, some writings, such as the Edwin Smith surgical papyrus, contained details about the brain and behavioral changes as a result of brain damage (Breasted, 1991). In it, a word for *brain* was used for the first time in writing, the convolutions of the cortex were likened to corrugations of metal slag, the meninges (membranes covering the brain) were described, and the behavioral results of brain injuries were given with reference to the side of the head that was injured and its effects on the limbs. A case study of a patient who sustained a major head injury was also described, in which the patient appeared to understand the attending physician, but the patient was speechless or voiceless, yet cried and wiped his tears while being questioned. The latter appears to be an example of Broca's aphasia (to be elaborated later in this chapter), in which comprehension is preserved but speech production is impaired.

Heart and Soul and the Ventricles

By at least 4,000 years ago, Egyptians had begun a sophisticated means of preparing dead bodies for an afterlife. The belief at that time was that the heart was the residence of not only the soul but also the intellect. Thus, upon death, only the lungs, liver, stomach, and intestines were accorded special treatment and placed in four separate canopic jars, whereas the brain was usually unceremoniously drained from the skull and discarded because it was not considered important. The heart was not removed from the body because it was considered so important in the afterlife. Finger (1994) has also noted that the debate about whether the intellect was in the heart or brain continued through the 16th century, as Shakespeare's Portia, in *The Merchant of Venice*, asks, "Tell me where is fancy bred, Or in the heart, or in the head?" It is also interesting to note that at least some of this heart bias still persists, as Finger has remarked that "people speak of having a 'broken heart,' giving 'heartfelt' thanks, or memorizing by 'heart' " (p. 15).

Roman physician Galen (130–200 CE) performed hundreds of autopsies, although predominately on animals. His view of the heart–mind connection was an amalgam between that of Plato, who thought the heart, brain, and gut interacted to form the intellect, and that of Aristotle, who was firmly cardiocentric, viewing the heart as the seat of intellectual reasoning, perception and sensation, and any related processes. Galen's view was that imagination, thinking, and memory all resided in the brain. However, he reinforced earlier Greek notions that the ventricles (the spaces in the brain containing cerebrospinal fluid) possessed "animal spirits," and no damage to the brain would result in symptoms unless the damage also directly affected the ventricles.

The Renaissance

The Renaissance (beginning in Italy from about the mid-1300s to late 1500s) produced a number of noted artists and polymaths like Leonardo da Vinci and Michelangelo. Both are known to have conducted hundreds of dissections of animal and human bodies, and it was during the Renaissance that speculations arose that challenged the prevailing notion that ventricles and accompanying animal spirits played a role in the human intellect. Interestingly, an American gynecologist, Frank Meshberger (1990), was one of the first to note that Michelangelo's *Creation of Adam*, painted on the ceiling of the Sistine Chapel, was an anatomically accurate picture of the human brain. Although certainly the picture remains a bit of a Rorschach card in that one may see what one wishes to see, there are a number of interesting coincidences; see Fig. 3.1.

First, God is portrayed in the frontal lobes, and as will be discussed in Chapter 4, they have been called a possible "center" of the brain or a metaphorical homunculus by neurologist Patricia Goldman-Rakic and her colleague Leung (2002). Neuropsychologist Muriel Lezak (1982) called them "the *heart* of all socially useful, personally enhancing, and creative abilities " (italics mine; p. 281). Neuroscientist Elkhonon Goldberg (2001, 2009) called

Fig. 3.1 Michelangelo's *Creation of Adam* (See color plate).
Source: Wikimedia Commons.

them "the essence of an individual, the personality core" (2001; p. 1). In addition, Meshberger made the interesting comment that Adam is not being given life by God, as Adam is obviously already alive in the painting, but Adam is being given his intellect, and God does so by extending his right arm through the prefrontal cortex (an area to be discussed in greater detail later). Second, another interesting coincidence, and well beyond that era's medical knowledge of the brain, is that God's left arm is above the shoulder of Eve. The coincidence is that Eve's head is centered in the parietal lobes (discussed in greater detail in Chapter 5), which is reputed to be where the human sense of self resides (e.g., Coolidge, 2014; Lou et al., 2004).

In 1664, English physician and Oxford University professor Thomas Willis published a critically important book, *Cerebri Anatome* (Latin for *cerebral anatomy*), on the brain and nervous system that profoundly influenced thinking about the brain and its functions for the next two centuries. Willis firmly argued that the complex cerebral convolutions of the human brain were responsible for the superior intelligence of humans compared with that of animals and that the cerebellum controlled involuntary actions, even in mammals. However, given the prevailing political and religious thinking of his time, Willis maintained the long-standing search for a soul, and this search pervaded and guided his anatomical work. Nevertheless, his careful brain and nervous system delineations influenced others in finding functional attributes of individual brain regions For further reading, see Finger's (1994) comprehensive early history of brain theories, which traces the neuroscientific cultures of Egypt, Greece, Rome, India, China, and others.

More Modern Brain Function Theories

In 1810, German neuroanatomist Franz Gall and his German physician and assistant Johann Spurzheim published an extensive book on the anatomy and physiology of the human nervous system and brain. They correctly made some important observations: differentiating between the cortex (upper layers of the brain) and its functions (e.g., thinking, feelings, and attitudes, predilections, etc.) and the subcortex (middle and lower levels), which they knew controlled vital life functions like respiration and heart rate. They also correctly surmised that there was an interaction between the two parts, but also some degree of independence. Gall and Spurzheim could be considered brain localizationists: specific regions of the brain perform specific functions.

Gall, however, incorrectly reasoned that the bumps and dips on the surface of the skull reflected underlying overdeveloped or underdeveloped functions such as one's intellect, hope, wit, causality, sense of time, morality, individuality, destructiveness, and many others. Spurzheim called Gall's theory *phrenology*, and the two actively promoted his ideas throughout Europe and America through public lectures and demonstrations, although the two had a falling out, and Spurzheim continued on his own, still promoting their "science." Many scientists were correctly appalled by the reading of head bumps, but one unintended consequence was that scientists became highly reluctant to assign specific functions to *any* part of the brain, and this reluctance persisted for nearly a century. As noted in Chapter 1, it is now known that specific groups of neurons may have dedicated functions (such as topographically organized neurons in part of the parietal lobe that differentiate between amounts and numbers of things), but any specific function always involves more than one region of the brain, and any one region of the brain is typically involved in more than one function.

The Advent of Modern Neuroscience and Clinical Neuropsychology

Russian neuropsychologist Alexander Luria (1902–1977) has been credited with founding clinical neuropsychology as a scientific discipline distinct from other medical and psychological endeavors. *Neuropsychology* is the study of brain–behavior relationships. The adjective *clinical* refers to the primary use of brain-damaged patients and animal models in investigations of the various functions of different regions and networks of the brain. At the age of 22, Luria, with a degree in medicine, began collaborating with Russian psychologist Lev Vygotsky, who was 28 years old. Together, with another Russian psychologist, Aleksei Leontiev (21 years old), they formed a formidable intellectual troika who studied and performed experiments assessing preserved and lost behavioral functions in brain-damaged patients. Many of these patients were World War I Russian soldiers. Later, Luria assessed World War II brain-damaged soldiers, and he also began investigating methods for their rehabilitation. He remained a prolific writer and investigator throughout his life; his tour de force was *Higher Cortical Functions in Man*, first published in 1962 (in English in 1966). His lifetime contributions to the field were many. He helped establish and create an interest in child neuropsychology, where he noted that brain damage had differing effects and outcomes in children compared to those in adults. He studied identical and fraternal twins to assess contributions of genetics and culture on behavior (modern behaviorial genetics). He studied and delineated different types of aphasia (speech and language disorders). He conducted a varying battery of neuropsychological tests on his patients that assessed an array of brain dysfunctions, although it is important to note that he stressed individual behavioral differences in brain-damaged patients such that a single, comprehensive battery of tests was difficult to attain. He was also particularly interested in frontal lobe–damaged patients and noted that the frontal lobes and prefrontal cortices are responsible for the most complex activities of all human behavior, that is, the ability to attend to present behaviors but also to imagine and integrate simulations of future behaviors. He viewed the frontal regions as a "superstructure" above all other cerebral areas and that they had the critical responsibility for the general regulation of all behavior. Finally, Luria (1966) astutely noted, "every higher mental function . . . is composed of many [neurological] links and depends for its performance on the combined working of many parts of the cerebral cortex, each of which has its own special role in the functional system as a whole" (p. 79). In this regard, Luria presaged later works by Anderson (2010, 2014), who also correctly noted that a single brain region may be involved in a multitude of functions, and any single complex human behavior has contributions from multiple brain regions and networks.

Ontogeny of the Brain

Rudiments of the human brain are visible within about 3 weeks of an egg's fertilization. There are at least four "building" principles in a brain's development: cell proliferation, cell differentiation, cell migration, and genetically directed cell death (apoptosis). As the zygote (fertilized egg at the two-cell stage) replicates by dividing itself repeatedly (cell proliferation), it begins to form a fried-egg shape. The thickening in the center is the formation of the embryonic disk. At week 3, a neural groove develops along the length of the embryonic disk, which will house the spinal cord and brain. In cell proliferation zones (called *patterning centers*), genes signal the development of specific forms and specific functions (cell differentiation). Genes then direct the cells to migrate to specific regions and, in cooperation with environmental signals, to form synapses. After week 3, the neural groove begins to close, and at one end forms holes in the neural tube that will later form the ventricles (spaces) of the brain. Eventually, the ventricles will be bathed by the steady production of cerebrospinal fluid, which cushions the brain, provides buoyancy, provides homeostatic functions for brain cells, and clears metabolic wastes. By week 7, the brain cells have proliferated, differentiated, and migrated enough to be clearly recognized as a brain (cerebrum), midbrain, subcortical structures (below the cortex), and the upper and lower spinal cord. The brain's regions are also genetically directed to make connections with other regions, and some regions are more neurally and intimately connected to each other than to other regions. For example, the front part of the brain (frontal lobes) is intimately connected to the upper middle part of the brain (parietal lobes) and to the lower (*inferior*, in brain terminology) and rear (*posterior*) part of the brain (temporal lobes). And, of course, the corpus callosum forms to transmit information between the left and right hemispheres.

Cell proliferation is nearly complete by week 20, but cell migration continues until about the 29th week. Cell differentiation and maturation (growth of axons, dendrites, and glial cells) continue after birth. Although there is the tendency to think that most brain cell death occurs in old age, and, indeed, by age 75 years a normal brain will weigh about 55% of what it was at 18 years of age (with individual variation, of course), brain neurons and their synapses begin a very precipitous decline after about 1 year of age and continue until about 10 years of age! This process of apoptosis eliminates about 50% of the brain's neurons and about two-thirds of the established synapses (also called *synaptic pruning*). Therefore, the brain is created more like a sculpture than a building, although initially it appears that the brain is built in layers on layers of neurons, and then unnecessary pieces are progressively removed over a relatively short period time. It is still unclear why particular brain cells and synapses are programmed for elimination. It is suspected, however, that it might be a case of "use it or lose it," where surviving neurons and synapses have been used, while those that have not been called on are eliminated. Thus, evolutionary processes provide a plethora of neurons and synapses for maximal behavioral flexibility but quickly eliminate unnecessary ones, which no doubt also minimizes metabolic requirements.

The Cerebral Hemispheres

The clearest visual landmark in a human brain is its two cerebral hemispheres, which together are referred to as the *cortex*, although strictly speaking only the upper six layers of the brain are considered the cortex. The cortex is about the thickness of an orange rind (2 to 4 mm, or about 1/8 in.). The left hemisphere is slightly bigger than the right in humans and in other great ape species (chimpanzees, bonobos, gorillas, and orangutans). In humans the potential for each hemisphere is different: the left hemisphere is genetically directed to process language, and the right hemisphere performs complimentary functions for language and some specific nonverbal and visuospatial functions. The two hemispheres are *highly interactive* in nearly all cognitive and emotional functions. The two hemispheres also appear to have greater neuroplasticity (modifications so the neurons can handle different tasks other than their original functions) before puberty than after puberty. However, if a left hemispherectomy (-*ectomy* means "removal") is performed on a postpubertal human, that person will be essentially mute, with limited language abilities. Before puberty, the right hemisphere is able to learn most but probably not all language functions, but after puberty this plasticity is lost.

The two hemispheres are separated by the longitudinal fissure (a deep cleft between the ridges of the brain) that runs from the anterior (toward the front) part of the brain to the posterior (toward the back) part of the brain. The two hemispheres communicate quickly by means of a commissure (communicative/connective brain tissue) called the *corpus callosum*, which is the largest commissure of the brain. The corpus callosum lies beneath the dorsal (toward the top) surface of the cortex. Split-brain studies, discussed later in this chapter, are performed by cutting the corpus callosum to reduce the frequency or severity of epileptic seizures.

There are four major ventricles (spaces) in the brain, which are filled with cerebrospinal fluid. The cerebrospinal fluid is created by the surfaces that line and form the ventricles. The fluid flows much like a river over the interior surfaces, and then it is reabsorbed. The flow is helped by cilia (tiny hair-like structures) lining the ventricles. If the flow is blocked, the resulting syndrome is hydrocephalus (*hydro* meaning "water," and *cephalus* meaning "brain"), which causes brain malformations and inhibited brain growth if it occurs in infants or young children. Shunting (siphoning) of this fluid reduces the fluid's pressure and ameliorates hydrocephalic symptoms. Interestingly, the ancient Greeks thought that the ventricles were the seats of all knowledge and filled with air, even though good anatomists could clearly see they were filled with a liquid.

The surface of a living brain is pinkish, reddish, and purple (veins and arteries) in color. The surface of dead brains is gray, which is what scientists have studied for ages, hence the name *gray matter*. Beneath the surface of the cortex it is substantially white, because the myelin sheath lining the axons of brain neurons is made up of fatty matter, hence the name *white matter*. The fat speeds up the transmission of nerve impulses. Glial cells called *oligodendrocytes* produce the myelin sheath for neuronal axons in the brain. Demyelinating diseases, such as multiple sclerosis (MS), interfere with axonal communication, and so motor movement slowing and disruption is often an initial symptom of MS, although later subtle memory problems may develop (Coolidge, Middleton, Griego, & Schmidt, 1996).

Brain Naming Systems

There are multiple naming systems for the brain and its regions. One of the earliest and still influential naming systems is Brodmann's areas (e.g., Brodmann's area 44, or BA 44). In 1909, German neuroanatomist Korbinian Brodmann classified the brain into 52 regions, roughly based on cytoarchitecture and function. Brodmann's system is still in use, but his numbering system was based on what region he chose to study next, rather than on areas of related functions or contiguous brain regions.

Another brain naming system is the presumed function of the region, such as motor cortex, somatosensory cortex, or primary visual cortex. In 1870, two German neuroanatomists, Fritsch and Hitzig, published a study in which they electrically stimulated the surface of the cortex of dogs and found muscle contractions on the opposite side being stimulated. They labeled that area *motor cortex,* based on its observed function, and references to the motor cortex are still being used.

Some brain areas are named after people, such as the Sylvian fissure, Broca's area, or Wernicke's area. A fourth brain naming system is the location of a region, such as the dorsolateral (top and side) prefrontal cortex (DLPFC) or ventromedial (toward the belly or middle) prefrontal cortex (VMPFC). Brain regions can also be named by the nature of their structure and their location, such as the superior (upper) temporal gyrus (a convolution's ridge) or intraparietal sulcus (a sulcus is a cleft between convolutions).

A sixth naming system is the gross (in the sense of what it looks like) anatomical name, such as the lunate sulcus because it looks like a sliver of the moon, or the hippocampus because it looked like a seahorse to early anatomists.

Lobes of the Cortex

The Frontal Lobes

The frontal lobes are the largest of the brain. The inferior (situated below something) convolutions of the frontal lobes are clearly demarcated by the lateral fissure (also known as the *Sylvian fissure, Sylvian sulcus,* or *lateral sulcus*). The posterior convolutions of the frontal lobes and the anterior potions of the parietal lobes are divided by the central sulcus (also called the *central fissure, central sulcus of Rolando,* or *Rolandic fissure;* see Fig. 3.2). Luigi Rolando was a 17th- to 18th-century Italian

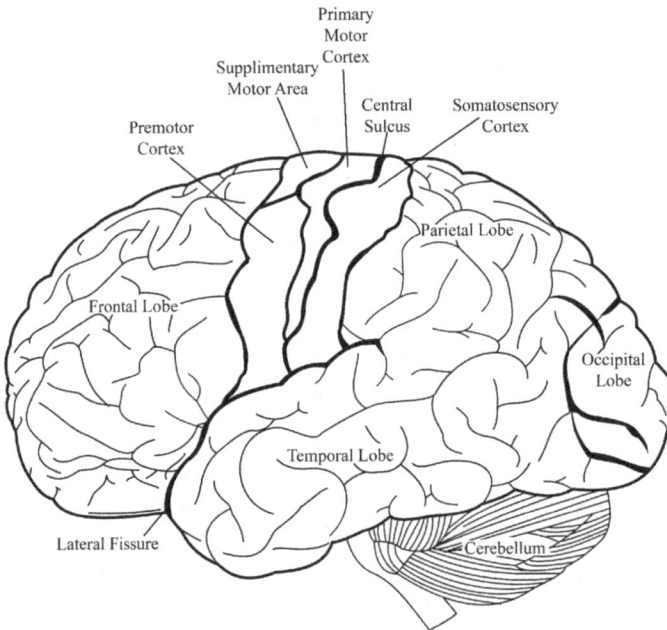

Fig. 3.2 Lobes and fissures of the brain.
Source: Adapted from public images by Tara Dieringer.

anatomist who may have been the first to note that there was a relatively fixed pattern to cerebral convolutions. He also observed that different areas appeared to have different functions, such as the motor and sensory convolutions, which are divided by the central sulcus. Unfortunately, his discoveries were partially obscured by his contemporaries Gall and Spurzheim and their phrenology claims. As noted previously, Fritsch and Hitzig (1870) later clearly established the function of the primary motor cortex. The right and left frontal lobes are also divided in half by the longitudinal fissure.

Primary Motor Cortex

As can be seen in Figs. 3.2 and 3.3, the primary motor cortex (BA 4) plans and executes motor movements for any muscle (e.g., hands, arms, limbs, tongue) by way of the spinal cord and its system of nerves (bundles of neurons). It does so in cooperation with many other areas of the brain, as no brain region acts alone in any function. Anteriorly and dorsally to the primary motor cortex is supplementary motor area (SMA; BA 6), which appears to generate plans, controls sequences of movements, guides one's posture whether sitting or moving, and aids in the coordination of bilateral (both sides) actions. Anterior to the primary motor cortex is the premotor cortex (also BA 6), which also plans and prepares movements and aids in the spatial guidance of movements.

Broca's Area

In the 1860s, French neurologist Paul Broca identified the left hemisphere as being critical to language. His claim went against the prevailing thought of the time that there were no areas of the brain with specific functions. As noted previously, that thinking persisted because of Gall's widely discredited phrenology. Broca, based on studies of many of his left hemisphere–damaged patients, proposed that the ability to speak was a left hemisphere function. An area of the left hemisphere is now called *Broca's area* (BA 44, or pars opercularis, and BA 45, or pars triangularis), and it is thought that damage to this region results in a specific neuropsychological syndrome known as Broca's aphasia (or expressive aphasia),

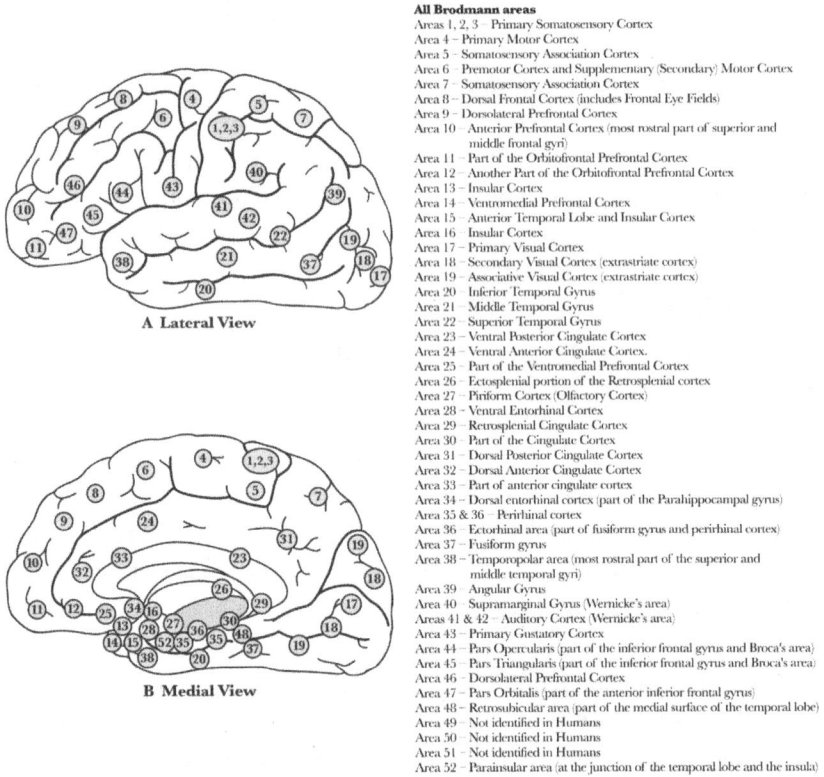

All Brodmann areas
Areas 1, 2, 3 – Primary Somatosensory Cortex
Area 4 – Primary Motor Cortex
Area 5 – Somatosensory Association Cortex
Area 6 – Premotor Cortex and Supplementary (Secondary) Motor Cortex
Area 7 – Somatosensory Association Cortex
Area 8 – Dorsal Frontal Cortex (includes Frontal Eye Fields)
Area 9 – Dorsolateral Prefrontal Cortex
Area 10 – Anterior Prefrontal Cortex (most rostral part of superior and
            middle frontal gyri)
Area 11 – Part of the Orbitofrontal Prefrontal Cortex
Area 12 – Another Part of the Orbitofrontal Prefrontal Cortex
Area 13 – Insular Cortex
Area 14 – Ventromedial Prefrontal Cortex
Area 15 – Anterior Temporal Lobe and Insular Cortex
Area 16 – Insular Cortex
Area 17 – Primary Visual Cortex
Area 18 – Secondary Visual Cortex (extrastriate cortex)
Area 19 – Associative Visual Cortex (extrastriate cortex)
Area 20 – Inferior Temporal Gyrus
Area 21 – Middle Temporal Gyrus
Area 22 – Superior Temporal Gyrus
Area 23 – Ventral Posterior Cingulate Cortex
Area 24 – Ventral Anterior Cingulate Cortex.
Area 25 – Part of the Ventromedial Prefrontal Cortex
Area 26 – Ectosplenial portion of the Retrosplenial cortex
Area 27 – Piriform Cortex (Olfactory Cortex)
Area 28 – Ventral Entorhinal Cortex
Area 29 – Retrosplenial Cingulate Cortex
Area 30 – Part of the Cingulate Cortex
Area 31 – Dorsal Posterior Cingulate Cortex
Area 32 – Dorsal Anterior Cingulate Cortex
Area 33 – Part of anterior cingulate cortex
Area 34 – Dorsal entorhinal cortex (part of the Parahippocampal gyrus)
Area 35 & 36 – Perirhinal cortex
Area 36 – Ectorhinal area (part of fusiform gyrus and perirhinal cortex)
Area 37 – Fusiform gyrus
Area 38 – Temporopolar area (most rostral part of the superior and
            middle temporal gyri)
Area 39 – Angular Gyrus
Area 40 – Supramarginal Gyrus (Wernicke's area)
Areas 41 & 42 – Auditory Cortex (Wernicke's area)
Area 43 – Primary Gustatory Cortex
Area 44 – Pars Opercularis (part of the inferior frontal gyrus and Broca's area)
Area 45 – Pars Triangularis (part of the inferior frontal gyrus and Broca's area)
Area 46 – Dorsolateral Prefrontal Cortex
Area 47 – Pars Orbitalis (part of the anterior inferior frontal gyrus)
Area 48 – Retrosubicular area (part of the medial surface of the temporal lobe)
Area 49 – Not identified in Humans
Area 50 – Not identified in Humans
Area 51 – Not identified in Humans
Area 52 – Parainsular area (at the junction of the temporal lobe and the insula)

Fig. 3.3 Brodmann's areas of the brain and functions (lateral and medial views).
Source: Holly R. Fischer, MFA.

in which a patient retains some ability to understand spoken language but either cannot speak or speaks "telegraphically," like omitting articles and conjunctions. However, recent studies (e.g., Tate, Herbet, Moritz-Gasser, Tate, & Duffau, 2014) using direct cortical stimulation of 165 patients undergoing brain surgery have shown that expressive aphasia only occurred when the ventral premotor cortex (BA 6) in both the left and right hemispheres was being inactivated and not when the classical Broca's area was inactivated. Tate et al.'s findings challenge the prevailing view that Broca's area is solely a speech production area. They found that Broca's area was involved in language production, including word retrieval, word-sound assembly, and sequential word processing, and concluded that it should be viewed as a supramodal (i.e., it transcends the senses) hierarchical language processor.

Prefrontal Cortex

Research for about the past five decades has consistently demonstrated the critical importance of the area anterior to the motor cortex, the prefrontal cortex (PFC), especially for the overall control and direction of behavior. It is commonly divided into the dorsolateral PFC (DLPFC; BA 6, 8, 9, 10, 46), orbitofrontal cortex (OFC; BA 11, 45, 47 and inferior portions of BA 24 and BA 32), and ventromedial PFC (VMPFC; BA 44). The PFC is critical to forming both short- and long-term goals and forming plans and strategies for attaining those goals. Interestingly, nearly four centuries ago, the French philosopher, mathematician, and scientist Descartes searched for a "center" of the brain, from which all thinking emanated. Ironically, he decided on the pineal gland because it was not a bilateral organ of the brain. It is called a "gland" because it secretes the hormone melatonin, which regulates sleep and both the circadian (24-hour cycles) and seasonal cycles. Interestingly, the pineal gland gets its name from it resemblance to a pine nut. Although, Descartes was wrong about the pineal gland's function, brain researchers such as Goldman-Rakic and Leung (2002) have wondered whether the PFC should be considered a "center" of the brain because of its role in decision-making and selected attention in both humans and other animals. The dorsolateral prefrontal circuit is generally associated with classic executive functions such as selective attention in spite of interference, organizing tasks in space and time, selective inhibition, response preparation, goal attainment, planning, and flexibility. The orbitofrontal prefrontal region is more closely connected to the limbic system of the brain and has been shown to be associated with the processing and regulation of emotions and decision-making associated with social behavior and interactions. The more central part of the OFC is called the *ventrolateral PFC* and the region toward the middle of the brain is called the *VMPFC* (see Fig. 3.4).

There is increasing evidence that the OFC may be an important part of a large and complex neural network (with major input from the hippocampus and other parts of the limbic system). This system forms a cognitive map that helps define the nature of a task or goal. It also forms decisions related which completes those tasks or goals. The OFC and this neural network may be particularly active when the decision to act depends on guesses or mental simulations of various outcomes of those actions. Thus, the OFC may be part of a system responsible for behavioral inhibition, calculating and signaling the consequences of errors, the attribution of errors to their appropriate

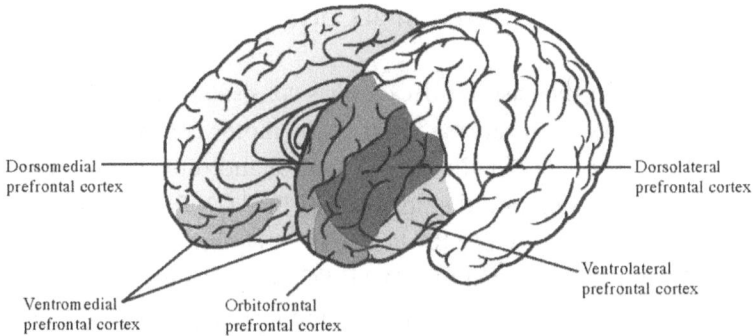

Fig. 3.4 Prefrontal cortices.
Source: Holly R. Fischer, MFA.

causes, and inferring the value of alternative actions when those actions are mentally simulated (and not simply based on prior outcomes). These associative cognitive maps, of course, would be critical in human evolution when alternative actions must be generated and a decision must be made about one of them, especially when those ancestral human types encountered novel problems with which they had no prior direct experience. It is again important to note that the OFC does not make these decisions in isolation but receives input from various other regions, such as the parietal and temporal lobes, subcortical structures, and the cerebellum.

For over 150 years it has been recognized that damage or intentional lesioning of the frontal lobes can result in profound changes to emotional and social functioning. The briefly popular, mid-20th-century frontal lobotomy (-*otomy* means "severing connections") rendered violent and highly anxious patients much less violent, much less anxious, and profoundly apathetic. These patients generally had no change in their IQ or memory functions, but they did become much less spontaneous, humorless, and blasé. Frontal lobotomies can be performed by severing connections within each frontal lobe (i.e., left and right frontal lobes) or severing the corpus callosum's connection between the two frontal lobes. Frontal lobotomies were performed mostly on severely psychotic or violent patients and sometimes on highly anxious patients whose behavior could not be controlled by medications. The operation became almost faddish in mental hospitals in the 1950s and early 1960s. However, as soon as the serious consequences of the operation became well known and as major new antipsychotic drugs became available, it was reserved for only the most difficult cases. The most often used defense was that it made the patient more "manageable." In 1941, Rose Marie Kennedy, the sister of the late US president John F. Kennedy, underwent a frontal lobotomy, at the age of 23. She had shown some intellectual difficulties early in life but reportedly became "difficult" and had violent mood swings in adolescence. Her father, Joseph P. Kennedy, then arranged for her lobotomy, without telling his wife. The surgery can only be described as crude, as the surgeons hammered a thin, ice pick–like knife into each section of her frontal lobes until she stopped speaking (she was only mildly tranquilized and was asked to say prayers, sing, or recite numbers). After the operation, it became clear that she was severely intellectually challenged, had palsied arms, was incontinent, and was unable to walk until much later in life (even then with a limp), and she could never speak clearly. She was institutionalized for the rest of her life (she lived 63 more years).

Portuguese neurosurgeon António Egas Moniz was awarded the Nobel Prize in 1949 for pioneering lobotomies, although there was and still is much controversy about his award because of the dire consequences for a majority of the lobotomized patients. See Fig. 3.5 for a replica of these lobotomy tools.

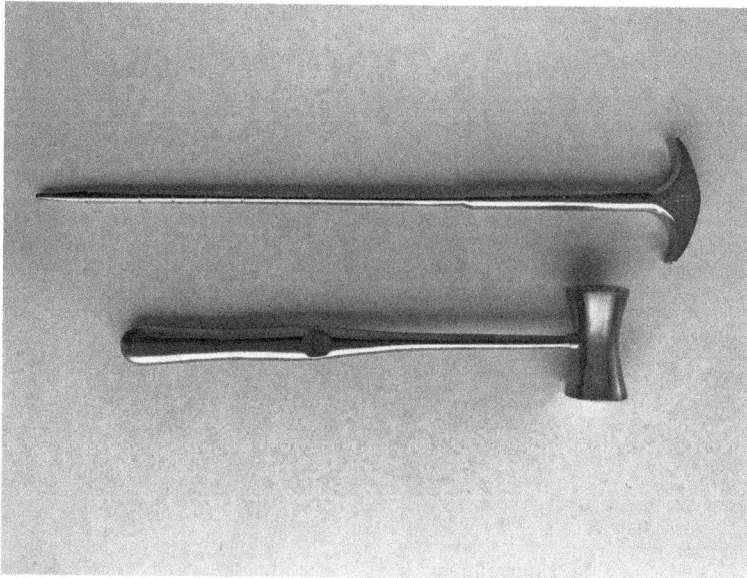

Fig. 3.5 Ice-pick lobotomy tools.
Source: Frederick L. Coolidge.

Cingulate Cortex

Inferior to the PFC and running anterior to posterior in the medial (middle) portions of the inferior cortex is the cingulate cortex, or cingulate gyrus (BA 23, 24, and BA 26, 29, 30, 31, 32). *Cingulum* means "belt" in Latin, and this region was given its name because it covers the corpus callosum (as can be seen in Fig. 3.6). The anterior portion of the cingulate cortex (BA 24) is known to be important to attention and selection of appropriate stimuli in external and internal environments and, perhaps, in maintaining consonance between short- and long-term goals. There is also some evidence that the anterior portion of the cingulate cortex is part of a brain network involved with the emotions of anger and disgust. The posterior portion of the cingulate gyrus (BA 29 and 30), the retrosplenial cortex (to be discussed shortly), is part of an important brain network along with the hippocampus, which is critical to making representations of one's environment. There is also recent evidence that abnormalities in the cingulate gyrus are frequently found in schizophrenic patients, a disease with severe social and some cognitive deficits, although other areas of the brain have also been implicated.

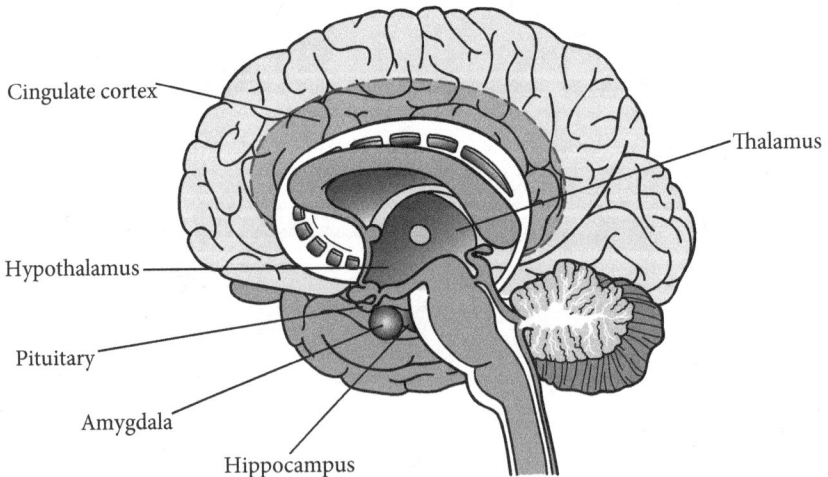

Fig. 3.6 Cingulate cortex and other midbrain structures.
Source: Adapted from Wikipedia Commons by Tara Dieringer.

Parietal Lobes

The parietal lobes (BA 1, 2, 3, 5, 7, 39, 40) have a relatively clear demarcation anteriorly, as they begin posterior to the central sulcus (see Figs. 3.3, 3.7, 5.2). This area includes the somatosensory cortex (BA 1, 2, 3), located in the postcentral gyrus, known to control and integrate the tactile senses of touch, including the senses of feeling, pain, temperature, movement, pressure, texture, shape, and vibration. The somatosensory cortex has intimate connections to the thalamus (to be discussed later), the latter of which is the gateway to and from the somatosensory regions. The parietal lobes are also critical in the manipulation of objects in visual space (visuospatial processing). Classic symptoms of damage to the parietal lobes are *apraxia*, an inability to execute intentional motor movements, and *agnosia*, an in ability to recognize one's own body parts. Again, however, the parietal lobes' neurons have been exapted for many higher cortical functions, as will be discussed later.

In terms of evolution, it is not surprising that sensory and motor pro cesses are located adjacent to each other in the brain, as early mammals and

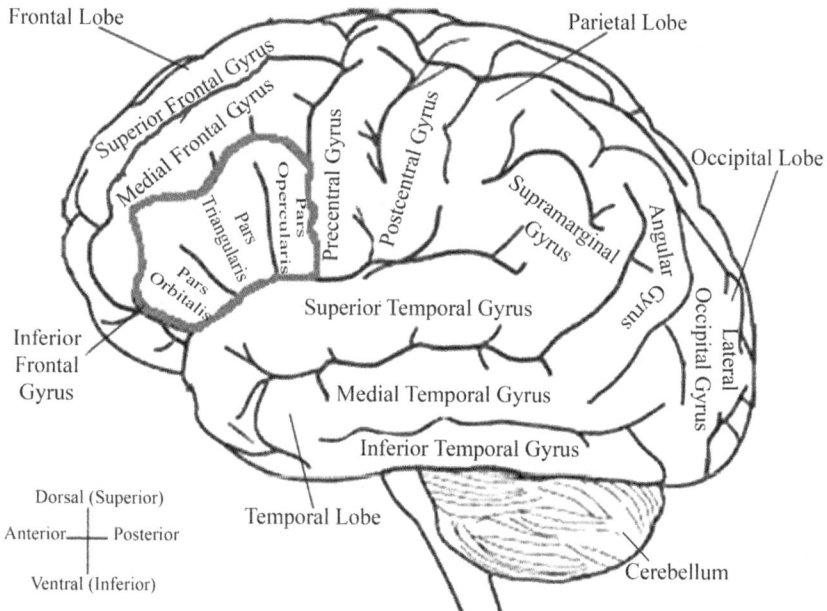

Fig. 3.7 Lobes and gyri of the brain.
Source: Holly R. Fischer, MFA.

then primates originally occupied a nocturnal, arboreal niche. These brain regions were critical to living and moving about successfully in trees. Even modern humans' brains have an inordinate representation for hands and feet, which would have been essential for early primates leaping among tree branches, beginning about 65 million years ago. It is also not surprising to find that modern humans' brains have exapted these same regions and adjacent regions for a sense of self (e.g., who we are and where are we going [metaphorically]).

There are two important regions in the inferior portions of the parietal lobes, the supramarginal gyrus (BA 40) and the angular gyrus (BA 39)—both areas have been implicated in language processing, coupled with other neural networks. The supramarginal gyrus (in both hemispheres) has been shown to be critical to the temporary storage of sounds (phonological loop) and of the linkage of sounds to meaning. It has been proposed that the supramarginal gyrus is the site of inner speech, although the idea remains debatable. The supramarginal gyrus is actively recruited when human subjects are asked to concentrate on the sound of words rather than on meaning. Neuroimaging studies have also confirmed its role in the maintenance of a verbal trace but not, perhaps, in the actual encoding of meaning. It also appears to be critical to verbal working memory tasks where continuous verbal information must be temporarily maintained and attended to despite interference.

The angular gyrus (BA 39), which sits posterior to the supramarginal gyrus, has long been known to have a role in mathematical calculation, and damage to this area may result in various arithmetic deficits known as *acalculia* or *dyscalculia*. It was also previously suspected to be the brain area where written words were transformed into an internal dialogue. More recent work has shown that the angular gyrus is critical not only to word reading and comprehension, but also to semantic processing, making sense of external and internal events, the manipulation of mental representations, and even out-of-body experiences (particularly the right angular gyrus). Current research also supports the hypothesis that the angular gyrus may be a multimodal hub for processing multiple sources of sensory information and making sense of that information. For well over a decade, the angular gyrus has been suspected of being involved with metaphor production and comprehension, although that role has been debated, but at the very least it is clearly activated when subjects are asked to compare concrete and abstract concepts (i.e., metaphor comprehension and production). Numerous studies have also confirmed its role in the verbal retrieval of numbers and their calculations, the verbal coding of numbers, the mediation of spatial representation of numbers, the recollection of personal memories (i.e., autobiographical memory), visuospatial navigation, and theory of mind (i.e., the ability to assume and understand the intentions and attitudes of other people, correctly or not). Interestingly, it has been implicated as part of the "default mode network" of the brain that becomes activated when people are resting and involved in their own thoughts without external engagement, but becomes deactivated when they are engaged in a goal-directed activity (e.g., Seghier, 2013).

Precuneus

Two areas of the medial and superior portions of the parietal lobe have also received much recent attention. The general area is identified as BA 7, both medially and laterally. The medial portion of the parietal lobe is known as the *precuneus* (see Fig. 3.3). Because of its location, it has been relatively difficult to study, but neuroimaging studies have been able to identify the precuneus as part of a network critical to autobiographical memories and prospective memory. The latter, of course, may have played a major role in human evolution as various problem-solving scenarios could be generated and internally debated, thus saving people from the dangers of actual trial and error (sometimes *fatal* trial and error). Again, it is important to note that the precuneus and other brain regions rarely act alone. Many different functions may be attributed to single regions, and single regions are almost always part of multiple neural networks. Laterally, BA 7 has long been known to be active, both in people and nonhuman primates, in determining the location of objects in physical space and determining the relation of those objects to one's body. Given humans' shared 65 million-year-old primate history (living in trees), these visuospatial abilities are thus ancient. Certainly, all intentionally moving animals also relied on brain regions that allowed them to "understand" their place in their environments, but this determination might have even been more critical for these early primates who spent their lives living, moving, mating, playing, and sleeping in trees.

Intraparietal Sulcus

Laterally and just below BA 7 is the intraparietal sulcus (IPS), which is again a shared structure between humans and nonhuman primates. However, the anterior portion of the IPS appears newly evolved in humans and has no clear homologue (which means "shared structures ") in apes or monkeys. Curiously, neuroimaging studies of neurons in the IPS have shown them to be dedicated to the representation and appreciation of numbers. The latter function is called *numerosity*, a capability shared by humans and many other animals. It has been hypothesized that there are at least two core processes in numerosity: subitization, the ability to differentiate between one, two, and three things, and set comparisons, the ability to differentiate between smaller sets of things and larger sets of things. Coolidge and Overmann (2012) and others have theorized that numerosity may have served as a rudimentary cognitive basis for abstraction, as the basic concepts of one, two, or three things can be applied to *any* things in the world, such as sticks, stones, bones, apples, and even mythical creatures like unicorns. Further, since human infants (as young as 8 months old) and monkeys demonstrate numerosity, then it is a function independent of language. It may have been useful evolutionarily to have a region of the brain (parietal lobes) that could immediately distinguish between one, two, and three things, be they predators or fruits, and pass that information on to other regions (frontal lobes) for a decision. In that same light, it may have been imminently useful to know that a tree with 50 fruits holds less than a tree with 100 fruits, although knowing the exact number of fruits in each tree would not be necessary (also see Kutter, Bostroem, Elger, Mormann, & Nieder, 2018).

Although brain regions' and networks' neurons have been exapted for purposes other than their original function, they may often also retain their original adaptive purposes. It certainly appears that the parietal lobes have undergone many such exaptations, such as taking on the functions of numerosity and higher number processing, sense of self, autobiographical memory, prospective memory (relating to the future), mental time travel, and the awareness that one's sense of time is relative (as one can recall past memories and manipulate them, as well as fabricate future memories, which, as noted before, is called *autonoesis*). However, it does appear that the parietal lobes retained many of their original functions, such as visuospatial relationships, directing limb and hand movements, and finger recognition, among others.

Paleoneurologist Emiliano Bruner's research over the past decade (2004, 2010; Bruner & Iriki, 2016) has shown that it is the expansion of the parietal lobes in modern humans (*Homo sapiens*) that best differentiates their brains from those of Neandertals. He has also demonstrated that the epicenter of this expansion may be the IPS and the precuneus (see Bruner & Iriki, 2016; Cavanna & Trimble, 2006, for additional information on the precuneus). We will return to some of the implications of this expansion in *Homo sapiens* in Chapter 5.

Retrosplenial Cortex

The retrosplenial cortex (RSC; BA 29 and 30) is located posterior to the splenium, the latter of which is the thickest and most posterior part of the corpus callosum. Along with the hippocampus, thalamus, precuneus, and medial temporal lobes, the RSC appears to have played a major synergistic role in the evolution of hominin cognition, particularly *Homo erectus*. Human and animal neuroimaging studies have shown that the RSC plays a central role in a network of brain regions for navigation, especially novel environments and spatial memories. This network has also been shown to have a role in episodic and autobiographical memories. However, its most critical function may reside in its ability to make transitions between an egocentric viewpoint (a view from one's self), known to be a precuneal or posterior parietal cortex function, and an allocentric viewpoint (viewpoint independent or a view from another person or place's perspective). According to Vann, Aggleton, and Maguire (2009), the place and grid cells of the hippocampus index locations contained within episodic or autobiographical memories, and then the RSC translates these indexes into egocentric and allocentric information such that a location in a memory may be viewed from a more specific viewpoint or other points of view. It is suspected that the RSC may also act as a short-term storage buffer while information is being translated. Many human neurophysiological studies have confirmed that the RSC is significantly activated by many kinds of spatial navigational tasks, including passive viewing of scenery, virtual-interactive spatial navigation, and active navigation of both new and highly familiar environments. The RSC is also highly active when topographical (maplike) information needs updating or for use of one's own motion to plan routes. Human and animal studies involving brain damage in these regions confirms the loss or major degradation of the aforementioned spatial abilities. As mentioned in Chapter 1, *Homo erectus* has been called a weed species, coming out of Africa many times over a million years or more, and it is known to have greatly expanded its territory compared to the australopithecines and habilines. The ability of *Homo erectus* to be successful in these endeavors no doubt depended on the natural selection of a network of brain regions that could reciprocally translate these egocentric and allocentric viewpoints.

Finally, Burgess, Becker, King, and O'Keefe (2001) have proposed that this translational RSC model may be related to imaginative or creative thinking for its basic ability to reconstruct scenes or imagine alternative scenes. Their model might help to account for the dramatic changes in technology from Mode 1 to Mode 2 stone tools. Thus, rather than the enhanced navigation abilities of *Homo erectus* and their bifacial handaxes being independently evolved behaviors, they may share a common neurological substrate.

Temporal Lobes

The temporal lobes (BA 20, 21, 22, 36, 37, 38, 41, 42) are clearly delineated anteriorly and superiorly by the lateral fissure. The temporal lobes lie inferior to the parietal lobes and lie anterior to the occipital lobes. They surround the hippocampus and the amygdala, important limbic system structures, which will be discussed shortly. The neurons of the temporal lobes appear to be specifically programmed to process language and sound interpretation functions, and, as such, regions of the temporal lobes are also referred to as *auditory cortex*. The temporal lobes thus play the major role in thinking, speech (internal and external), and memory.

The temporal lobes are also often referred to by its three gyri. The posterior portion of the superior temporal gyrus (BA 22) is also called *Wernicke's area* (although there is no universal and definitive region that constitutes Wernicke's area), after Carl Wernicke, a German neuroanatomist. In the late 1800s, Wernicke observed that not all aphasias result from damage to Broca's area, and Wernicke was able to demonstrate that this region was responsible for the understanding of speech. Damage to this area results in Wernicke's aphasia, in which the understanding of speech is impaired but some ability to speak is retained.

Another region within the posterior part of the superior temporal gyrus is the planum temporale. It is typically larger in the left hemisphere than in the right hemisphere, and this asymmetry has been claimed to be more predominant in chimpanzees. In humans, symmetry in the left and right hemispheres in this region has been found to be associated with some learning and reading disabilities. The planum temporale has been shown to have a role in music, particularly the perception of pitch and harmony.

The superior temporal gyrus also contains an area (BA 41, 42) known as the *transverse temporal gyrus*, which is responsible for hearing and basic sound processing, and it receives input directly from the cochlea (inner ear). This area is also known as the *primary auditory cortex*.

Medial Temporal Gyrus

The medial (middle) temporal gyrus, like the other two gyri of the temporal lobes, is involved with aspects of language processing and memory. It has also been implicated in judgments of the attractiveness of faces, which is not surprising given that it is adjacent and superior to the inferior temporal gyrus. It appears to be part of a ventral temporal network involved in the recognition of objects and faces. It may also be part of the brain's default mode network. The entorhinal cortex is in the medial temporal region, which appears to serve as an interface with the hippocampus. Thus, its primary purposes may be the same as those of the hippocampus: spatial navigation and verbal memory formation.

Inferior Temporal Gyrus: Fusiform Gyrus

The inferior temporal gyrus also processes language, particularly word and number recognition. One specific area, the fusiform gyrus (BA 36, 37), has neurons dedicated to the recognition of faces and objects, in humans, other great apes, and monkeys. In humans, the fusiform gyrus also selectively responds to word forms. Interestingly, there is a large degree of lateralization of function between the left and right hemispheres for this region in humans. Written words activate the left fusiform gyrus, which is labeled the *visual word form area* (VWFA), and faces activate the right fusiform gyrus (fusiform face area, or FFA). The VWFA is activated regardless of the writing system used, although it has been shown that nearly all writing systems employ only three basic forms.

The inferior temporal gyri are also connected to the extrastriate cortex (BA 18 and 19), a critical part of the primary visual cortex (BA 17) in the occipital lobes. The extrastriate cortex is connected to the posterior portion of the parietal lobes. Damage to the left VWFA produces *alexia*, a complete inability to read words.

Epilepsy

A disorder frequently associated with damage to the temporal lobes is epilepsy. The disorder has appeared frequently in the history of neuropsychology, as much has been learned about the brain through methods to diagnose and treat epilepsy. It is a seizure disorder, which means that brain waves begin to act abnormally, and adjacent brain regions and similar regions in the other hemisphere sometimes mimic the abnormal waves, thus spreading the seizure. Seizures interfere with consciousness, and sometimes they are accompanied by convulsions. There are many causes for epilepsy, for example, head injuries and tumors, but most epilepsy is said to be *idiopathic*, which means of unknown origin, although it is suspected that most cases of idiopathic epilepsy originate at birth from a lack of oxygen (anoxia) to the brain. It may be that the temporal lobes are most susceptible to anoxia or are the first of the lobes to be sensitive to anoxia. If the initial onset of symptoms (called an *aura*) in epilepsy is associated with dreamlike states, altered states of consciousness, and dissociative states (e.g., out-of-body experiences, feeling robotic, etc.), then an informal diagnosis of temporal lobe epilepsy may be indicated. If the site of the seizures can be localized through an electroencephalographic (EEG) evaluation or other neurological measures or procedures, then this site is called the focal *epileptogenic* area. As noted previously, the identification and treatment of epilepsy have played a major role in the history of understanding brain structures and functions.

Occipital Lobes

The occipital lobes (BA 17, 18, 19; see Figs. 3.2, 3.3, and 3.7) sit posteriorly to the parietal lobes and temporal lobes, and their chief function appears to be primary visual recognition and processing. There are no clear demarcations of the occipital lobes from the parietal or temporal lobes, although a structure known as the *lunate sulcus* more clearly demarcates the occipital lobes from the parietal and temporal lobes in apes and monkeys. It does appear that the lunate sulcus has moved posteriorly in the evolution of the cortex of *Homo sapiens*, suggesting a diminished role of the primary visual cortex in modern humans compared to that in other primates. It has also been surmised that Neandertals had larger primary visual cortices than those of *Homo sapiens* living at the same time, which may suggest a more important role for the sense of vision in Neandertals.

Damage to the occipital lobes, if diffuse (widespread) enough, can result in blindness because of the *contrecoup* effect (on the side opposite that was hit). Pugilistic blindness, sometimes seen in boxers, results from being hit in the front of the head so often or so severely that the occipital lobes are damaged. Interestingly, in some of these cases, patients may deny being blind, and they provide poorly masked excuses for not being able to see, such as "it is too dark to see." Providing imaginative excuses to mask testing errors is called *confabulation*. It is not known to what extent confusion plays a role in the mixing of imagination and memory or what other motivations may be involved. Visual neglect or visual inattention may also result from damage to the occipital lobes (see further discussion later).

As noted previously, the primary visual cortex (at the extreme posterior end of the cortex) is BA 17. The adjacent areas, BA 18 and 19, are called the *extrastriate cortex,* and one of its functions appears to be the perception of other people's body parts and their intended movements. It also processes "higher-level" visual information like color, hue, object recognition, and movement. The flow of visual information from the occipital lobes to other brain regions appears to depend on the type of information. Information about the location of objects and their motion is transmitted to more posterior portions of the parietal lobes, and it is labeled the *dorsal stream* (the "where" stream of information). Damage to the network of the dorsal stream usually results in deficits in spatial orientation and the detection of motion and in difficulties with visual tracking. Information about the nature of an object, its color, shape, or form is called the *ventral stream* (the "what" stream). The ventral stream, along with temporal regions, is involved in word recognition and meaning, reading, attention, learning, and memory.

Insular Cortex

The insular cortex is a distinct, oval-shaped group of gyri, located deep within the Sylvian fissure. It is hidden by parts of the frontal and temporal lobes. It has reciprocal connections to the brainstem, thalamus, amygdala, basal ganglia, frontal, temporal, and parietal lobes, and all sensorimotor association areas. Its anterior portion, the anterior insular cortex (AIC), in conjunction with the anterior cingulate cortex (ACC) and the prefrontal cortices, plays an important role in emotional awareness and sensing physiological states of the body (e.g., temperature, pain, etc.). Damage to the AIC may result in *alexithymia*, in which an individual has deficits in emotional awareness; alexithymia is seen frequently in schizoid personality–disordered individuals and in autism spectrum disorders (e.g., Coolidge, Estey, Segal, & Marle, 2013). The AIC, ACC, and DLPFC have some large spindle-shaped cells called *von Economo neurons* (VENs), whose chief function appears to be transmitting information quickly across distant areas of the brain. As VENs are primarily found in larger-brained primates, like the great apes, it is suspected that they may have evolved relatively recently, perhaps within the last 20 million years. They are also found in larger-brained and more intelligent mammals like whales. Thus, one hypothesis is that they evolved independently (called *convergent* evolution) in great apes and whales but serve similar purposes: rapid transmission of information across big brains, with some role in the processing or maintenance of social behavior.

Limbic System

The limbic system, originally conceived of as the emotional processing center of the brain, consists of a number of brain structures or regions, but its status as a system is somewhat arbitrary, because it is partially based on location and partially on general functions. In general, the limbic system is involved in the processing of emotions and the formation of memories including verbal, visual, and olfaction, and visuospatial locations. However, with regard to the processing of emotions, it is recognized that emotions are highly complex, thus they are unlikely to have a single brain circuitry or simple neuronal basis. That these two major functions, emotion and memory formation, are entwined in a common neural substrate also suggests an intimate relationship between the two, and empirical research strongly supports the

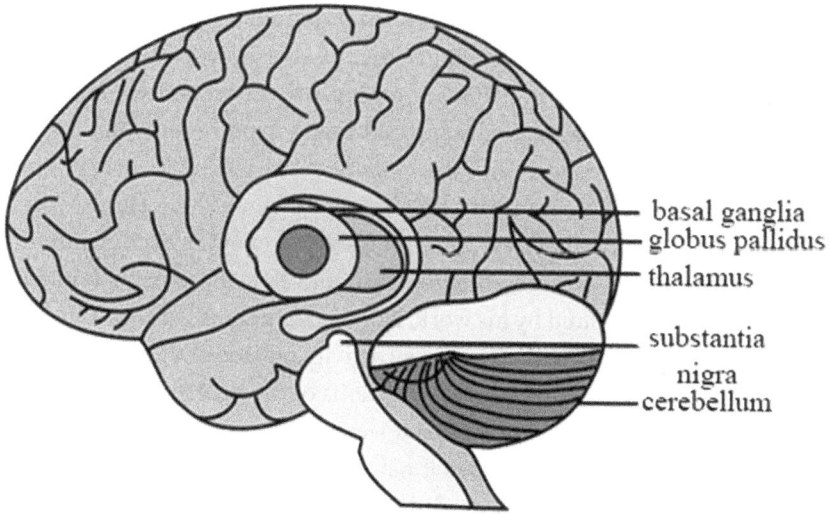

Fig. 3.8 Basal ganglia and other structures.
Source: Adapted from Wikipedia Commons by Tara Dieringer.

role emotions play in memory formation. The limbic structures are housed ventrally (the underside or belly of the cerebrum) and medially within the cortex and temporal lobes. Limbic system structures most often include the hippocampus, amygdala, cingulate cortex, hypothalamus, mammillary bodies, nucleus accumbens, parahippocampal gyrus, OFC, and parts of the basal ganglia, including the striatum, globus pallidus, and substantia nigra. They are all bilateral structures (appearing in both hemispheres). Limbic structures are phylogenetically older than the cortex, and they are far more prominent in nonmammalian brains. This chapter will address in some detail only three of these structures—the hippocampus, amygdala, and basal ganglia. See Figs. 3.8 and 8.1.

Hippocampus

The hippocampus is a bilateral, horseshoe-shaped structure whose first known purpose was the memorization of spatial locations in both humans and animals. All mammals have a well-developed hippocampus compared to that in fish, reptiles, and birds. There are homologous structures in the latter animals, and those structures also appear to be involved in spatial cognition.

The German philosopher Immanuel Kant proposed long ago that the perception of space was a mental ability that existed independent of experience. In the 1960s and early 1970s, an American-British neuroscientist, John O'Keefe (and his colleague), substantiated Kant's hypothesis by finding that rats had neurons called *place cells* that became active when rats were in a particular place in their environment (O'Keefe & Burgess, 2005). He also found that these place cells would rearrange themselves in new environments, thus creating new maps of those environments, and that those maps remained stable over time. Inspired by his work, Norwegian neuroscientists May-Britt and Edvard Moser found additional cell activity outside of the hippocampus in rats and mice with connections to portions of the medial temporal lobes. They called these cells *grid cells*, which were able to approximate distances, forming a neuronal basis for spatial navigation. For their combined work, O'Keefe and the Mosers were awarded the 2014 Nobel Prize in Physiology or Medicine.

The hippocampus is also critical to the formation of declarative memories (also known as *semantic memory, verbal memory,* or *explicit memory*) but not to the formation of procedural memories (like stone knapping or juggling). Damage to the hippocampus can result in anterograde amnesia, in which new verbal memories cannot be transferred to long-term memory. Damage to the hippocampus can also result in the loss of memories up to about 2 or 3 years past, so newly learned memories appear to stay vulnerable to forgetting during this period. Because the hippocampus is considered part of a network for emotional processing as well, the connection between emotional states and memory is not inadvertent. Humans remember things that have an emotional valence and tend to forget things that do not. Thus, the emotional valence of an event (highly dependent on one's prior beliefs or attitudes) will determine whether something is memorized or not. If the event is exceptionally emotionally arousing, such as natural disasters, war, death, and other catastrophes, a person will have a very difficult time forgetting the event, and some will develop posttraumatic stress disorder as a result.

Amygdalae

These bilateral, almond-shaped structures (*amygdala* is the singular form), on the anterior tips of the hippocampus, play a well-researched role in emotional processing, particularly fear and rage responses. It has also been suggested that the amygdalae's chief function is to determine what an external stimulus is and what should be done about it. Although the amygdalae have often been touted as the neural home of emotions, an overwhelming majority of this research has been devoted to only two emotions, rage and fear. Amygdalae are also highly complex and interconnected with other regions of the brain, and so some suggest that they are not a single structural entity, nor should they be considered a single functional unit. Its basal nucleus receives input from all sensory systems. Its central and medial nuclei are involved with the output of innate emotions, and its cortical nuclei's input comes from the olfactory bulb and the olfactory cortex (also called *piriform cortex*). This olfactory input into the amygdalae's cortical nuclei gives a solid neuronal basis to the saying "the smell of fear," as Mujica-Parodi et al. (2009) have empirically demonstrated that humans can detect airborne chemical substances from emotionally stressed novice skydivers, and functional magnetic resonance imaging (fMRI) revealed activation of the amygdalae in the human detectors. The amygdalae also have receptors for at least five different neurotransmitters, receptors for at least two different hormones, and peptide receptors for oxytocin, opioids, and others.

As noted earlier, because emotions play a key role in memory formation, amygdalae play an important role in learning and memory as well. The amygdalae and other limbic structures are phylogenetically much older than the surrounding cortex, and the amygdalae and other limbic structures are often more prominent features in the brains of reptiles. In studies of mammals and primates, amygdalectomies result in a "taming effect" and electrical stimulation of the amygdala can instill rage reactions. Rage reactions have even been demonstrated in mice against much larger natural predators. Amygdalectomized humans become apathetic (as was the case in H.M., as will be discussed in Chapter 8) and show little spontaneity, creativity, or affective expression.

American neuroscientist Joseph LeDoux (1996) has noted that the amygdalae receive their sensory input from two different pathways, one fast ("the low road") and one slow ("the high road"). The fast pathway involves recognition from the visual cortex of, for example, a fearful animal, such as a snake. This recognition is sent to the thalamus and directly to the amygdalae, which may then elicit an immediate motor response, for example, "jump." In this circuit, autonomic (involuntary) responses also occur, such as increased heart rate and blood pressure, without full conscious recognition that the object is a harmless corn snake. Meanwhile, the visual cortex has simultaneously sent that information about the snake to higher levels of cortical processing, providing much greater resolution about the snake ("Oh, it's not a poisonous coral snake, it's a harmless corn snake"), and that information is still sent to the amygdalae for assessment, but it is a much slower process. Evolutionarily it seems that this dual system of analysis allows both fast recognition and accurate recognition, albeit at different processing speeds.

Basal Ganglia

The basal ganglia are a collection of subcortical neurons whose most clear function appears to be the control of movements. A section of these ganglia, the substantia nigra, manufactures dopamine, a neurotransmitter responsible for the control of movement. In the initial phase of Parkinson's disease, the neurons in the substantia nigra are destroyed, initially resulting in tremors of a finger, hand, or foot. Muscles often become rigid and stiff, especially facial muscles, resulting in a masklike face. Eventually, the ability to walk becomes impaired, and patients may move with a shuffling gait. In the later stages of Parkinson's disease, the loss of the surrounding brain tissue will result in dementia (memory problems and a loss of intellect) and, in some patients, psychotic symptoms. The initial dopamine production problem can often be ameliorated by levodopa (L-dopa), a drug that mimics dopamine. Hazy, Frank, and O'Reilly (2006) investigated the role of the basal ganglia in controlling actively maintained task-relevant information in the PFC. They found that the basal ganglia provide "Go" or "No Go" input into the PFC, which is learned through trial and error. Further, they concluded that this input helps support more abstract executive functions of the PFC.

Other Brain Structures

Cerebellum

The cerebellum (a diminutive in Latin meaning "little brain" or "lesser brain") has long been associated with the acquisition, maintenance, and smooth timing and execution of motor movements (see Fig. 3.9). In 1993, Japanese neuroscientist Masao Ito speculated that the cerebellum might control thoughts or ideas just like it controlled motor movements, and a host of

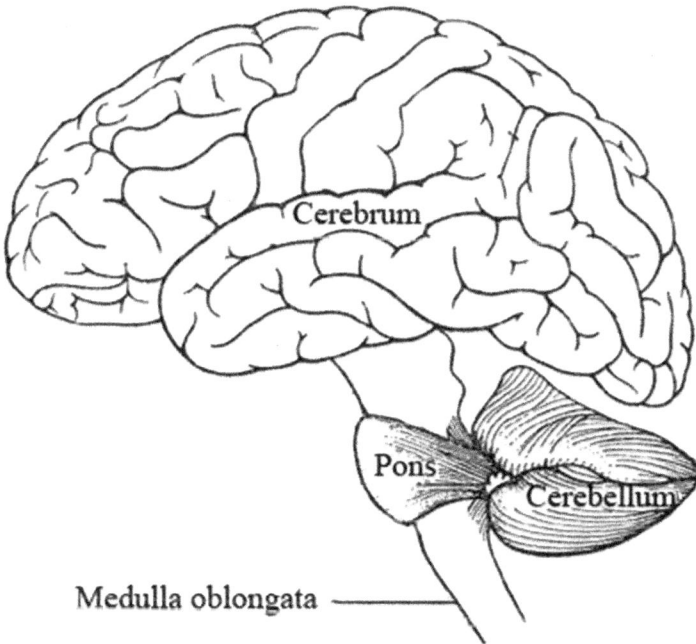

Fig. 3.9 Cerebrum, cerebellum, pons, and medulla oblongata.
Source: Adapted from Wikimedia Commons by Tara Dieringer.

neurophysiological studies over the past decade have substantiated his hypothesis. The cerebellum has now been implicated in a wide variety of cognitive functions, including insight, intuition, creativity and innovation, novel problem-solving, language, affective word meanings, aspects of grammar, metalinguistic skills (awareness of the subtleties of other's speech), verbal and visuospatial working memory, verbal fluency, reading, writing, and the rate, force, and rhythm of actions. In the great apes and monkeys, just as in humans, the anterior portion of the cerebellum controls motor movements, but it appears that the posterior portion of the cerebellum has expanded in *Homo sapiens* and has been exapted for these higher-order cognitive functions. The adaptation and exaptation of the cerebellum will be dealt with in greater detail in Chapter 7.

It has also been suggested that the evolutionary expansion of the cerebellum in humans and its exaptation for higher cognitive functions did not occur in isolation (mosaic evolution; the independent evolution of brain regions) but in tandem with the evolution of the prefrontal cortices and motor systems (concerted evolution). Thus, natural selection may not have been acting on individual brain regions but on functional and interconnected brain systems. As the first technologies may have been stone-tool making about 3.4 million years ago, the brain areas required to make stone tools, such as the PFC, motor systems, and cerebellum, may have all been selected for in concert, and it may help to explain how subsequent higher-level cognitive functions like language may have been able to exapt systems already in place for other functions. Interestingly, recent research suggests that Neandertals may have had a smaller cerebellum than that in modern *Homo sapiens* (although the former had about 10% bigger brains than the latter). This discussion will continue in Chapter 7.

The Fasciculi

As evidence of the concerted evolution of areas of the brain, the fasciculi (singular = *fasciculus*) serve as major neural transmission highways between brain areas naturally selected for their conjoined functions. The *arcuate fasciculus* primarily connects the temporal lobes to the frontal and prefrontal cortices. These connections demonstrate the intimate relationship between the functions of the temporal lobes, such as the meanings of words, thoughts, and ideas to the frontal areas involved in sequencing, reasoning, and other higher-level processing such as decision-making. The *superior longitudinal fasciculus* connects the frontal areas to all other lobes of the brain. The latter connections reflect all of the critical functions of the parietal, temporal, and occipital lobes to the functions of the frontal lobes. The *inferior longitudinal fasciculus* primarily connects the temporal lobes and occipital lobes. There are also fasciculi (e.g., uncinate) that connect the various vital functions of the brainstem to the upper cortices. See Fig. 3.10.

Brainstem

The brainstem forms the upper part of the spinal cord. The upper part of the brainstem, the pons, receives sensory and motor input from the body and sends it to both halves of the cerebellum. The pons is also the site where information from the body crosses over to be processed by the opposite side of the brain; that is, information from the left half of the body is sent to the

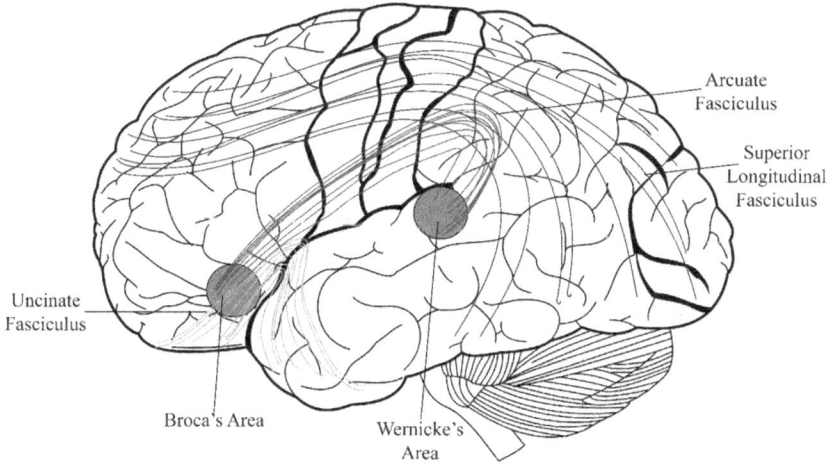

Fig. 3.10 Fasciculi.
Source: Adapted from public images by Tara Dieringer.

right side of the brain for processing and vice versa. The reticular formation is housed at the same level as the pons, and it is one of the oldest phylogenetic areas of the brain. It has connections to the thalamus, hypothalamus, cortex, and cerebellum. It appears to regulate many basic automatic functions such as sleep, eating, sex, alertness, attention, and motivation. Below the pons is the medulla oblongata, which processes the flow of information between the body and the brain. The medulla also controls most involuntary muscle movements and activities of the glands, and it controls vital functions such as heart rate, breathing, and blood pressure. Any major damage to this area usually results in death.

Thalamus

This egg-shaped structure (*thalamus* means "inner room" in Greek) at the top of the spinal cord is a gateway to the cortex for sensory systems. It has reciprocal loops with all regions of the cortex. Brainstem structures also innervate thalamic functions, and it has reciprocal loops with these structures as well. Specific sections of the thalamus have more specific functions; one of the most important functions is attention, regulated by the pulvinar nucleus in the posterior part of the thalamus.

Hypothalamus

This structure sits below the thalamus, and it has projections to the PFC, amygdala, and spinal cord. It also connects to the pituitary gland beneath it. The hypothalamus regulates the autonomic nervous system, the endocrine and hormonal systems, and the body's general homeostasis. The hypothalamus is also reciprocally affected by hormones circulating in the blood; thus there is a high degree of reciprocal interaction between the hypothalamus, the pituitary gland, endocrine/adrenal glands, and the hormones they regulate and produce. The hypothalamus, the pituitary gland, endocrine/adrenal glands form the "HPA axis" which is the body's stress response system.

The Senses

Touch and Handedness

Each hand has ipsilateral (same side) and contralateral (opposite side) connections to the two cerebral hemispheres. However, the contralateral connections between hands and hemispheres are much stronger than the ipsilateral connections; thus, the right hand is an instrument of the left hemisphere, and the left hand is an instrument of the right hemisphere. For example, after a major right hemisphere stroke (a disruption of the blood supply to an area of the brain), a patient will generally be paralyzed on the left side of the body and will be unable to gain control of a hand through the ipsilateral connections from the brain.

Approximately 90 to 95% of people favor their right hand, either by their stated preference or by their performance. For the other 5 to 10% of people, they most often claim they are left-handed or ambidextrous. In neuropsychological testing, handedness is most often assessed by the speed at which a person can tap their index finger on each hand. Interestingly, nearly half of those who claim to be left-handed will finger-tap faster with their right index finger than with their left, whereas nearly all people who claim to be right-handed tap faster with their right index finger. Because people who claim to be left-handed have such variable neuropsychological testing results, they are sometimes referred to as non–right-handed people.

It is interesting to note that right-hand preference appears to be largely a human trait. Only in chimpanzees has a weak right-hand preference been demonstrated, and typically only in captivity, which may indicate that the chimps are emulating their human caretakers (e.g., McGrew & Marchant, 1997). Also, nearly all right-handed people have speech located in their left hemisphere, and about 70% of non–right-handed people also have speech in their left hemisphere. It is thought that less than about 10% of non–right-handed people are "mirror" left-handed people; that is, they have speech in the right hemisphere and favor their left hand.

Although right-handedness appears to be a human trait and many major language functions are located predominately in the left hemisphere in humans, there is left hemisphere localization for vocalizations in a wide variety of animals, including birds, amphibians, and mammals. Broca's area in the left hemisphere, noted for its role in speech production in modern humans, appears to be enlarged in *Homo habilis*, about 2.5 million years ago, but not in australopithecines, about 3 million years ago (Holloway, 1983). Assessment of specific brain areas in these ancient hominins is difficult, given that cranial morphology must be estimated strictly from internal fossil skull features and the brains themselves rarely, if ever, fossilize. However, anthropologists tend to be in some agreement that *Homo habilis*, perhaps more than the australopithecines, may have been able to vocalize and thus may have had some sort of a prototypic language.

The causal relationship of language lateralization and hand preference is highly debatable. Corballis (2003) has argued that vocalization appears to have an early, left hemisphere localization (100 million years ago or more), while handedness emerged about 2 million years ago, probably as manual gestures were incorporated into the communications among *Homo habilis*. He speculates that as manual gestures were removed as a requirement of language, rapid advances in technology may have occurred in *Homo sapiens*. Corballis has also noted that Broca's area in chimps has been found to possess mirror neurons, which produce excitation when a chimp witnesses another chimp manually gesturing. Thus, if vocalization had a very early evolutionary left hemisphere localization, and manual gestures served as a basis for language, then mirror neurons may have aided in the acquisition of manual gestures and reinforced a right-hand bias for gesturing. Corballis offers further evidence by Toth (1985), who examined stone tools from 1.4

to 1.9 million years ago (more than likely associated with *Homo erectus*). He found that they were produced by right-handers more than by left-handers at a ratio of about 1.3 to 1, in contrast to the modern ratio of right-hand preference to non–right-handed preference of about 9 to 1. Curiously, Uomini (2009) found a near absence of left-handed knappers in more recent hominins such as *Homo heidelbergensis*, Neandertals, and Upper Paleolithic *Homo sapiens*. The latter finding may suggest that there was an increasing preference for right-handedness across our evolutionary past, although whether this bias occurred through vocalization's ancient left hemisphere localization, genetic drift, cultural biases, or a combination of factors is uncertain. I favor the hypothesis by Fitch and Braccini (2013) that *Homo sapiens*' strong preference for the right hand and left hemisphere localization for speech are probably independently derived traits. They suggest that human handedness preference probably comes from more ancient vertebrate perceptual asymmetries. Thus, rather than viewing human handedness as a "driving force," they see it as perhaps a "misplaced" focus and a red herring in the search for direct causal links to language's lateralization in the brain.

Ears and Hearing

The connections of the ears to the cerebral hemispheres follow the same pattern as that for the hands. Each ear has ipsilateral and contralateral connections to the hemispheres. However, the contralateral connections between ears and hemispheres are much stronger than the ipsilateral connections; thus, the right ear is an instrument of the left hemisphere, and the left ear is an instrument of the right hemisphere. In neuropsychological research and assessment, a dichotic listening paradigm has been predominately employed in which different words or short lists of words are played simultaneously to each ear. Each ear and hemisphere hears the words distinctly, as they are not jumbled together; however, the participant or patient can only reply with a single word at a time. Right-handed people show a strong preference for first reporting words from their right ears, because a word played in the right ear goes directly to the left hemisphere where speech resides (because the contralateral connections are stronger than the ipsilateral connections). A word played to the left ear goes directly to the right hemisphere; however, to be spoken, the word must be transferred to the left hemisphere via the corpus callosum. This small delay thus favors words played to the right ear.

Eyes and Vision

Each eye is connected to both hemispheres. Contralateral and ipsilateral connections are of equal strength. The left half of the left eye has an ipsilateral connection to the left hemisphere, and the right half of the left eye has a contralateral connection to the right hemisphere. The same is true of the right eye: The right half of the right eye has an ipsilateral connection to the right hemisphere, and the left half of the right eye has a contralateral connection to the left hemisphere. The left half of each eye views the right field of vision (right visual half-field), and the right half of each eye views the left field of vision (left visual half-field). Thus, if a participant or patient is focusing straight ahead (in neuropsychological assessment, patients focus on a star or circle directly in the middle of their field of vision) and a picture of a key is shown in their left field of view and a picture of a ball is shown in their right field of view (but very briefly), right-handed participants will predominantly report "ball, key" rather than "key, ball." This occurs because what is shown in the right field of vision will appear in the left half of each eye, and the left halves of the eyes are connected directly to the left hemisphere, where speech resides. The picture of the key (in the left field of vision) will be perceived in the right half of each eye, which are connected to the right hemisphere. The knowledge of the key must them be transferred from the right hemisphere over the corpus callosum to the left hemisphere in order to be spoken. See Fig. 3.11.

Split-Brain Studies

Neuropsychologist Roger Sperry won the Nobel Prize for Physiology or Medicine in 1981 for his work with intractable epileptic patients, that is, those whose seizures could not be controlled by medication. Sperry had noticed that seizures would often begin in one hemisphere (the epileptogenic focus) and spread to the other hemisphere. This occurs because the two hemispheres are so synchronous in their various functions that they mimic the seizure activity of the other hemisphere. Sperry found that severing the corpus callosum in patients reduced seizure activity significantly. He also found that accompanying behavioral repercussions of this commissurotomy appeared to be minimal. However, subsequent testing of these split-brain patients offered great insight into each hemisphere's varying roles in

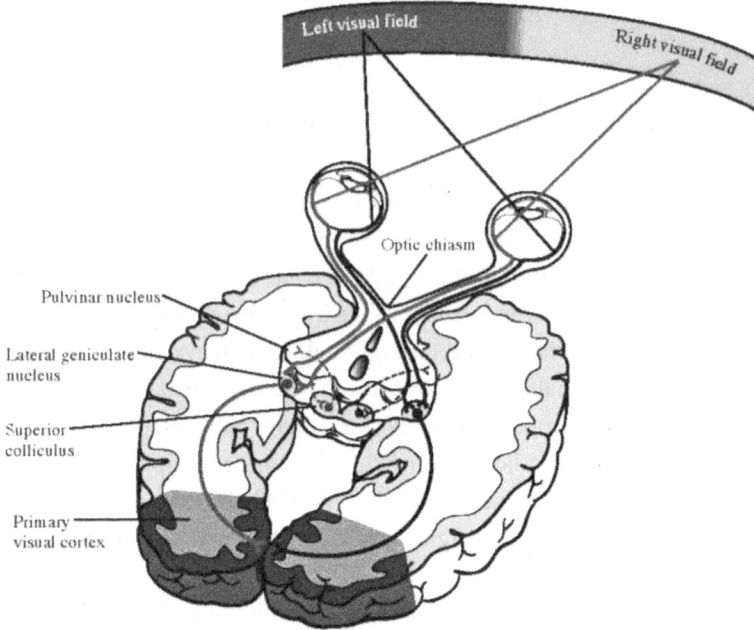

Fig. 3.11 Organization of the eyes and their visual fields.
Source: Holly R. Fischer, MFA.

perception, understanding, and speech. Sperry and his students found the left hemisphere was better at analytical, sequential, and linguistic tasks, and the right was better at some nonverbal and spatial tasks. Interestingly, Sperry and his students did not always agree on the implications of their findings about the extent of language functions of the right hemisphere. Their work set the stage for the popularization of left and right hemisphere differences, and overlooked in the subsequent hype was the extent to which the two hemispheres work synchronously on virtually all tasks.

Smell and Taste

The posterior portion of the human nose is lined with sensory cells (olfactory epithelium) that detect odors, based on about 350 functional olfactory receptor genes (rats have about 1,000). The axons from these sensory cells pass through the cribriform plate, a perforated bone in the anterior and inferior part of the skull. From there, the axons project directly to the olfactory bulbs, which lie directly and bilaterally under the human brain. The olfactory bulb has connections to the limbic system, particularly the hippocampus, amygdalae, piriform cortex, and the orbitofrontal cortex. Thus, the olfactory neural circuitry is the only sensory modality that is directly integrated into cerebral regions involved in higher cognitive functions, especially where memory and emotionality are intertwined; thus it may be argued that olfaction is a higher cognitive function and serves a much more sophisticated role than simply smell. Animal studies of olfactory bulbectomies have been shown to be a causative factor in reduced cellular immunity. Human studies have also shown that olfactory impairments (anosmia) herald autoimmune diseases, such as lupus, and many neurodegenerative diseases, including Parkinson's disease, Alzheimer's disease, and other types of pathology like depression. Interestingly, congenitally anosmic people also have been shown to have significantly more social and reproductive problems than congenitally blind or deaf people (although that finding requires replication). Bastir et al. (2011) have noted that the cribriform plate is about 12% larger in Homo sapiens than in Neandertals. As human olfactory bulb size has been shown to be correlated with olfactory performance, that is, odor threshold detection and odor identification, Levulis and Coolidge (2016) have suggested that there may have been significant behavioral consequences resulting from enlargement of the cribriform plate: It may have given Homo sapiens better general immunity, resistance to autoimmune diseases, and enhanced social, emotional, and sexual/mating functions.

Taste and smell, of course, are intimately related, as anyone with a stuffy nose will usually report a reduced sense of taste. Taste buds on the human tongue are sensitive to at least five distinct tastes: salty, sweet, sour, bitter, and umami (a brothy or meaty taste). Axons from the taste buds join with a facial nerve (the seventh cranial nerve) to connect to a region of the brainstem and then to the thalamus. The gustatory information from the thalamus is connected to the primary gustatory cortex (BA 43; part of the insula cortex), and that information is connected to secondary processing areas in the OFC, where taste and smell are finally integrated and evaluated (e.g., good, bad, noxious, etc.). Interestingly, sommeliers (wine experts) show similar initial brain activation in BA 43 in response to wine as do non-experts, but the sommeliers had increased activation in the insula and orbitofrontal and dorsolateral cortices, presumably because of enhanced cortical processing involved in decision-making and selection.

There may be evolutionary reasons for the five basic tastes. First, detection of bitterness and sourness is far more sensitive than that of saltiness, sweetness, or umami. It is suspected that bitter and sour were far more useful in avoiding poisonous substances than the other three taste dimensions. Also, smell was likely more important evolutionarily in poison detection than touch, taste, vision, or audition, because unlike touch and taste, smell could aid in avoiding direct contact with a poisonous substance, and vision and sound were unlikely to convey any useful information about the nature of poisonous substances.

Final Thoughts

Brains are the most complicated of all animate or inanimate systems. As such, simplistic brain models (even the ones presented in this chapter) and myths about the brain abound in human culture. Even brain damage treatments and brain rehabilitation programs are often fraught with pseudoscientific approaches and essentially useless or worthless help (other than, perhaps, a placebo effect). In sorting out the latter, I stress two thoughts. First, does the treatment have a clearly established scientific basis? It is not sufficient to hypothesize simply that improved cerebellar performance might enhance higher cognitive functions or improve psychological welfare. It should be demonstrated empirically each step of the way how that might be true. The defense of many promoters of such programs (often people with good intentions) is that many well-established approaches in science, like psychotherapy, have not been fully investigated and some causative brain–behavior links have not yet been completely established. That fact leads me to my second recommendation, which is that one must then go to great lengths to demonstrate that the treatment or program is not simply an enhanced placebo effect. Also, one must watch for claims that seem too good to be true (because they most often are not true) or are essentially entertaining but harmless. An example of the latter is the pervasive brain myth (totally without any scientific basis) that humans use only 10% of their brains. The claim is completely specious and silly; however, if one surmised upon hearing it that the other 90% could be used to learn Spanish or Hindi, or take up tango lessons, then the hypothesis is still false, but harmless or even beneficial.

Summary

1. The transition from life to consciousness may have begun with the ability to learn associatively, that is, classical and operant conditioning, during the Cambrian period, beginning about 545 million years ago.

2. There are at least four cell-building principles of the brain: cell proliferation, cell differentiation, cell migration, and programmed cell death (apoptosis).

3. There are multiple naming systems for the brain: (1) Brodmann's areas (e.g., BA 44), (2) presumed function (e.g., primary auditory cortex), (3) people's names (e.g., Broca's area), (4) regional location (e.g., superior parietal lobe), (5) location and nature of the structure (e.g., anterior intraparietal sulcus), and (6) gross anatomical name (e.g., lunate sulcus).

4. There are four major lobes of the brain: (1) frontal, (2) parietal, (3) temporal, and (4) occipital.

5. The limbic system, known to be important to emotional processing, memory, and movement, is an arbitrary group of brain regions, consisting of the hippocampus, amygdalae, basal ganglia, and others.

6. The fasciculi comprise major neural transmission connections, which reflect the concerted evolution of various brain regions.

7. Brains are the most complicated animate or inanimate systems in the universe.

4

The Frontal Lobes

As noted in Chapter 1, human brains have been influenced by at least two major evolutionary lineages. First, the divergent evolution of arboreal mammals from terrestrial animals led to a dramatic expansion of the cortex such that it has been labeled the *neocortex*. The expansion of the surface of the brain was apparently selected for its ability to be flexible, allowing quickly changing behaviors to a variety of changing environmental conditions and challenges. The output of the neocortex in mammals also enabled its axons to connect to midbrain structures but also bypass them and connect to all levels of the central nervous system. The actions of the primary motor and sensory regions of the brain became more specialized, allowing for a more detailed representation of the world, its objects, and players. Mammalian behavior began to extend beyond stereotypic movements (fixed action patterns) and anticipate and form plans of actions in advance. Making accurate representations of the world meant selecting for areas of the brain to form internal images; these areas included primarily posterior cortical structures (adjacent to primary visual cortices) and areas adjacent to these primary areas, called *association cortex,* which allowed for more complex cognitive processing. Finally, areas were selected that could form and anticipate actions (frontal areas of the brain; see Fig. 4.1). These changes to mammalian brains allowed both more accurate fixed action patterns (instincts) and much greater behavior flexibility than that of other animals. Thus, mammalian brain evolution has been typified by both mosaic evolution (natural selection for independent brain structures) and concerted evolution (selection for structures which perform complimentary functions, as noted in Chapter 3). Again, one powerful example of the concerted evolution of structures is the fasciculi, which coordinate the operations of various areas of the brain (also noted in Chapter 3; see Fig. 3.9).

The second major influence on brains was the divergent evolution of primates from the mammalian lineage. As noted earlier, primates were small, nocturnal, insect-eating, tree-dwelling animals that evolved from a common ancestor, beginning about 65 million years ago. Primates also ate leaves and

Evolutionary Neuropsychology. Frederick L. Coolidge, Oxford University Press (2020).
© Oxford University Press.
DOI: 10.1093/oso/9780190940942.001.0001

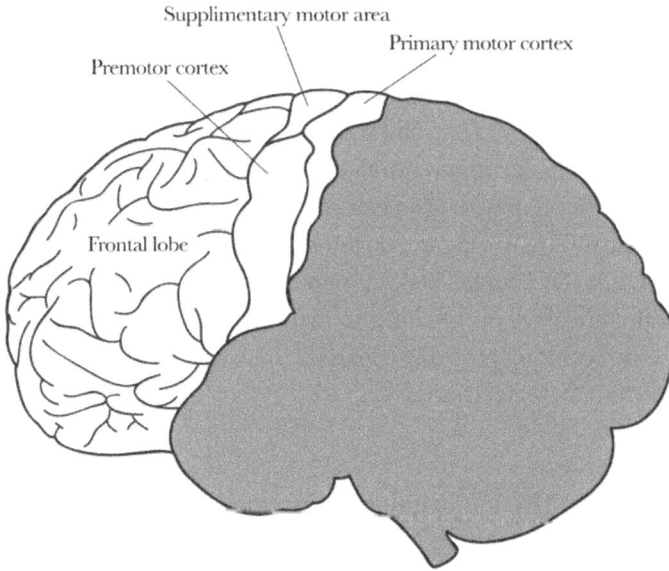

Fig. 4.1 Frontal lobes.
Source: Holly R. Fischer, MFA.

more highly nutritious fruit, and they probably competed for the latter with birds, reptiles, and other mammals, as well as other primates. However, to compete successfully, they needed highly accurate sensory and motor systems, particularly visual systems for recognizing ripened fruits and motor systems for locomotion through tree branches. In addition to visual and motor systems, audition was also a necessity. By coordinating their foraging activities with vocalizations, primate socialization became more sophisticated than that of other animals, and areas of the brain were selected for that could produce and comprehend (memorize) sounds, undoubtedly the inchoate beginnings of language. As a function of a more complex, socially oriented brain, it is often argued that human brains are simply scaled-up primate brains (which is only partly true). Yet primates have disproportionately larger brains for their body size (called the *encephalization quotient* or EQ) than mammals. Neuroscientist Suzana Herculano-Houzel and her colleagues (2014), for example, have argued that the metabolic requirements for human brains are the same as for other nonhuman primates relative to brain size, and that the human brain has the same ratio of neurons in its cortex to the rest of the brain as found in other primates. She notes that what is extraordinary about human brains is the sheer number of neurons (which she empirically estimates at 86 billion) and at least as many glial cells. Interestingly, Herculano-Houzel (2012) suspects that the development of cooking overcame the energy limitations of a raw diet; the great apes, for example, spend from about one-quarter to one-half of their waking hours chewing. Not having to expend this amount of energy on eating may have led to the development of an extraordinary number of neurons and glial cells in human brains. Cooking also provided immediate nutritional benefits over raw foods. Of course, cooking requires a fire, which was unlikely for the australopithecines or habilines to make since they still mostly dwelled in trees. Thus, cooking probably began with the advent of *Homo erectus* and their transition to full terrestrial life.

The Uniqueness of Human Brains

Besides a remarkable number of cells (neurons and glial) in human brains, there are a number of other unique features. Buckner and Krienen (2013) have argued that the human cerebral cortex has proportionately greater amounts of association cortex relative to primary sensory or motor areas compared to all other animal brains. In the late 1800s, German neuroanatomist Paul Flechsig helped identify these regions of the cortex as areas that serve more complex, higher-level cognitive processes. To Buckner and Krienen, an expanded association cortex suggests that more of the human brain is devoted to conceptual and higher-level cognitive-processing operations than to simply perceptual and motor processes. They also favor a "noncanonical" (i.e., lesser known brain pathways) brain circuitry for association cortex in addition to traditional canonical (i.e., common, specific, or well-known pathways) circuitry. In the traditional view of brain networks, dense areas between regions aid the hierarchical flow of information, for example, where sensory information is linked to motor responses, in a kind of feedforward and feedback connectivity. In non-canonical brain circuitry, association areas may conform to this type of connectivity but also appear to connect to other association areas without a consistent hierarchical organization. Further, this type of organization may result in many parallel functional systems that connect to all parts of the cerebrum. The gist of this association cortical organization is that it appears to be greatly expanded in hominin evolution, which may have been driven by top-down control and by internally mediated world representations.

Adaptations of the Frontal Lobes

As stated earlier, brains are for moving. Living things without true brains, like comb jellies, sponges, and plants, are generally not considered to move intentionally. Sure, young sponges may float about but then "root" and become sessile. Comb jellies can move about by means of cilia, and their rudimentary nervous system can react to some stimuli (e.g., temperature) by cilia, but again, they move mostly unintentionally. But at about 545 million years ago, the first flatworms moved purposefully, and their rudimentary nervous systems required some specialized cells to do so—in other words, brains. Also, as noted earlier, it became a selective advantage if their brains (or rudimentary eyes) were in the front of their bodies (at the "head") to better direct their movements. Thus, one of the first adaptations of brains, particularly the frontal or rostral part of the brain, was for the direction of purposeful movements. With the evolution of reptiles, amphibians, and, later, mammals at about 200 million years ago, limbs, especially forelimbs, became specialized for grasping and grabbing. Again, it appears that the forward portions of the brain, the frontal lobes, directed these grasping and grabbing movements, and this adaptation was well in place by the advent of the first primates about 65 million years ago.

Mirror Neurons

If neurons became adapted for simple but purposeful movement, then it could be argued that some of these neurons became exaptations for grasping and grabbing. This is what appears to have happened in evolution, be it an adaptation or an exaptation, but clearly the latter (grasping) followed the former (movement) evolutionarily. Beginning in the 1980s, Italian neurophysiologist Giacomo Rizzolatti and his colleagues explored the single neuron firings in the brains of macaque monkeys in a brain area F5, which is roughly the same as Broca's area in humans (BA 44 and 45). They found that single neurons would fire when the monkey grasped an object, and the same neurons would fire when the monkeys observed another animal (monkeys or humans) grasping an object. These neurons were labeled *mirror neurons.* Rizzolatti and his colleagues also found them in the inferior parietal lobes (Rizzolatti & Craighero, 2004). Through additional research they found that mirror neurons would fire when mouth and facial gestures were observed or mimicked. Subsequent human research found mirror-like properties for neurons in other areas of the brain of macaque monkeys, such as some regions of the temporal lobes (Keysers & Gazzola, 2010).

While provocative and interesting, speculations arose hypothesizing that mirror neurons and mirror neuron systems were responsible for a wide variety of cognitive and social behaviors and problems, such as the ability to understand the attitudes and feelings of others (called *theory of mind* [ToM]) and autism spectrum disorders. Some claims were made that mirror neurons might be to the cognitive sciences what DNA's discovery was to biology. Many of these claims were exaggerated, theoretically baseless, or, not yet, empirically justified. Cognitive scientist Gregory Hickok (2009, 2014) has summarized these criticisms and notes that central to most of the criticisms is the lack of theoretical basis for how mirror neurons might influence cognitive and emotional functions and empirical evidence justifying these and other claims.

In 2006, two UCLA neuroscientists (Iacoboni & Dapretto, 2006) hypothesized that mirror neuron system dysfunction was a core deficit in autism spectrum disorders (it has been called the "broken mirror" hypothesis). However, subsequent criticism as well as research with autistic children and adults appears to show that autistic people are able to understand their own actions and the intentions of other people's actions and that mirror neuron deficiencies or impairments in autistic people have not been reliably and empirically demonstrated. According to Hickok (2014), mirror neurons remain an interesting phenomenon and have helped in understanding how the brain works, but they must be viewed as a part of a "broad class of sensorimotor cells that participate in controlling action using a variety of sensory inputs and through a hierarchy of circuits" (p. 240). He feels that some founding mirror neuron theorists overemphasized the motor system's role in cognition and that they ignored sensory systems, other multimodal systems, and higher-level cognitive processing networks.

Prefrontal Cortex

The exaptation of the neurons in the prefrontal cortex (PFC) is one of the major evolutionary advances in hominin brains. The PFC resides in the anterior, ventral, and medial portions of the frontal lobes (BA 8, 9, 10, 11, 12, 44, 45, 46, 47; see Fig. 3.4). Generally, references to the PFC exclude the primary motor cortex (BA 4), premotor cortex, and supplementary cortices (BA 6). While most neuroscientists agree that, compared to other primates, the frontal lobes are not extraordinarily large relative to other parts of the brain, those claims belie the fact that there is greater white matter volume (mylenated neurons) in the PFC, there is greater gyrification, a unique neuronal cytoarchitecture, and it has substantial connections to other cortical and subcortical structures (e.g., all other lobes, cingulate cortex, basal ganglia, and the cerebellum; Rilling, 2014). Those claims also belie the fact that the human PFC performs more complex, higher cognitive and abstract processing.

Executive Functions of the Prefrontal Cortex

On September 13, 1848, Phineas Gage, age 25 and foreman of a railroad building crew in Vermont, accidentally dropped a 3 ft 7 in. long and 13¼ lb tapered iron tamping rod onto a dynamite charge. The rod entered the upper orbit of his left eye and exited through his median skull line above his frontal lobes (Fig. 4.2). According to a published medical account by Dr. J. M. Harlow (1848), Gage's crew carried him to an oxcart, and he was driven approximately a mile to a hotel. With some help, he walked up a flight of stairs to a bedroom that became his hospital room for the next 3 months. Amazingly, given the dearth of medical treatments for infections, Gage survived the accident and lived nearly 12 more years. He apparently died of unremitting convulsive seizures (*status epilepticus*) on May 21, 1860, in San Francisco. Harlow (1868) subsequently wrote a follow-up story of Gage, his travels to South America, and his brief stint in a circus, posing with his iron tamping rod.

Gage's story became celebrated and controversial throughout the latter half of the 19th century. In part, Gage's accident became a test case for phrenology and brain localization theories. If Gage had lost specific abilities, then it would be evidence for brain function localization and, thus, by association, evidence for phrenology. When a balance was struck between legitimate localization theorists and holists in the early 20th century, Gage's specific disabilities became less pertinent. Since then, Gage's case has more often become an example of personality changes that may result from frontal lobe damage, and it has passed into near mythology by modern textbook writers that after the accident "his friends and acquaintances said he was 'no longer Gage'"

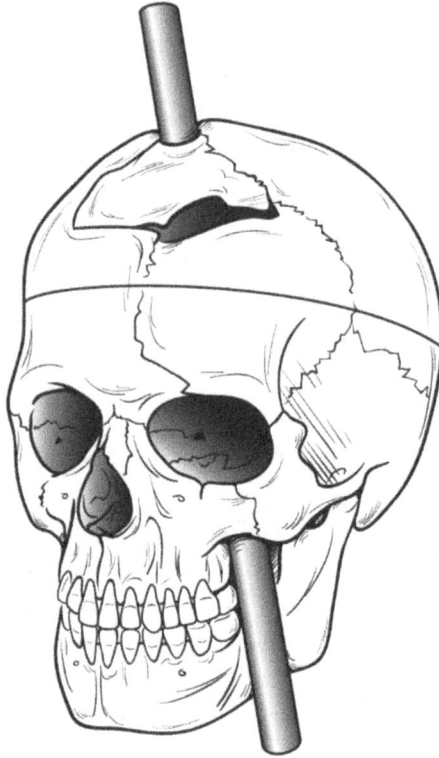

Fig. 4.2 Phineas Gage.
Source: Holly R. Fischer, MFA.

(Harlow, 1868, p. 277). However, what has been missed in the telling of Gage's story is that this quote has been taken out of context, and what Harlow was employing may have been the first use of the executive functions metaphor for the PFC. The critical phrase comes from a paragraph by Harlow (1868) describing a meeting between him and Gage about 8 months after the accident. The passage follows:

> Previous to his injury, though untrained in the schools, he possessed a well-balanced mind, and was looked upon by those who knew him as a shrewd business man, very energetic and persistent in executing all of his plans of operation. *In this regard* [italics mine] his mind was radically changed, so decidedly that his friends and acquaintances said he was "no longer Gage."

It seems relatively clear that Harlow was referring to Gage's shrewd business sense and executive planning functions, as well as his energy and persistence in completing his tasks. It is prophetic that Harlow used the wording "persistent in executing all his plans of operation," since this heralds findings over 100 years later that the PFC is involved in behaviors now known as *executive functions*, and especially attentional control(e.g., Lezak, 1982; Luria, 1962/1966). As such, Gage's story is not only a prelude to the hypothesis that brain damage may result in disruption of executive functions and in personality changes, it is also the first clearly written description that the frontal lobes serve as a kind of metaphorical executive: making decisions, forming plans and goals, organizing and devising strategies to attain those goals. These were probably the very skills that made Phineas Gage such a valuable foreman of his construction crew. Perhaps it was not simply a personality change that denied him regaining his job after the accident, rather, primarily it may have been this loss of his executive abilities. The problem with subsequent descriptions of Gage's accident in the psychological and neuroscience literature is that Harlow associated the phrase "no longer Gage" with Gage's loss of his shrewd business sense and his executive abilities and *not* with generic personality changes. This repeated inaccuracy in the literature results in two problems. First, Gage did have some personality changes, but that is not what Harlow was referring to when he wrote "he was no longer Gage." Evidence for personality changes in Gage's behavior resides elsewhere in Harlow's two articles, including symptoms of capriciousness, childishness, and telling fantastical stories. A second major problem with these accounts is that Harlow was describing perhaps an even more important sequela of Gage's accident: the loss of executive functions of the frontal lobes.

Over 100 years later, Lezak (1982) summarized the executive functions as follows: (1) the capacity to form goals, (2) planning, (3) carrying out activities, and (4) effective performance. According to Harlow's (1848, 1868) writings, Gage was deficient in all of these criteria. It appeared that he had no long-term goals, as he changed jobs frequently, including his aforementioned sad stint in a circus sideshow appearing with his iron rod as the man who survived a horrible blow to his brain. He had severe trouble forming plans; Harlow wrote that he made numerous plans in quick succession but abandoned them even more quickly. It is also clear that Gage had trouble executing any reasonable plan and was ineffective in carrying out even mundane tasks.

Broca's Area Revisited

At the outset of this discussion, it is important to keep in mind Anderson's (2010, 2014) hypothesis, which has substantial empirical support, that one region of the human brain may be involved in multiple functions (e.g., the angular gyrus may be involved in over 15 different cognitive and emotional functions; see Chapter 5), and to remember its corollary hypothesis, that a cognitive or emotional function always involves multiple brain regions. It is difficult, then, to overcome the standard bias that Broca's area (in the left hemisphere) is *the* speech production region of the brain. In an empirical study of 165 patients undergoing neurosurgery for cancer of the glial cells (glioma), Tate and his colleagues (2014) conducted direct cortical stimulation while giving the patients various cognitive tasks including motoric, sensory, counting, and a broad variety of language skills. As the surface of the brain has no pain receptors, the patients remained awake during surgery and testing. Direct cortical stimulation of a specific region of the brain allows neuroscientists to determine what areas are required to conduct a specific function. One of their most provocative findings was that speech output was not affected by stimulation of Broca's area; localization appeared to be a ventral premotor cortex function in both the left and right hemispheres. Broca's area did appear to be involved in higher-level aspects of language function such as semantic, phonological, and sequential processing. Their results support previous studies that have found Broca's area to be a higher-level multimodal processor and challenge the standard notion that Broca's area is solely involved in speech output.

Dorsolateral Prefrontal Cortex

The PFC evolved in different phases. These phases initially reflect changes to the brains of mammals such that increases in the size of the cortex relative to the rest of the brain came to be described by the term *neocortex*, as noted previously. Later changes to the PFC are reflective of the history and behavioral nature of primates. The first nocturnal tree-dwelling primates had to search for appropriate food items, favoring fruits, whose nutritional value over leaves allowed for further brain expansion. Improved food selection especially relied on the PFC, as did the motor coordination of head and hand movements required to eat fruit, which is a novel behavior compared to the feeding behaviors of reptiles and birds. The dorsolateral PFC (DLPFC) lies at the end of the dorsal stream of information from the visual cortex, so it makes complete sense that it might be involved not only in the visual selection of food choices, such as determining whether fruits were ripe enough to eat, but also in determining what motor movements were required to obtain food. Further, successful and unsuccessful movements had to be retained (memorized), to enable efficient and flexible behaviors in the future. Thus, it can be seen that the PFC evolve to act in concert with other areas of the brain for visual recognition (occipital regions), spatial locations (parietal and hippocampal regions), and motor movements (premotor cortex and supplementary areas), in addition to using auditory sounds (temporal regions) to coordinate feeding activities, avoid prey, and find mates.

One major unique contribution of the PFC is its ability to generate goals, either tangible or abstract. As will be discussed in greater detail in Chapter 5, my colleague cognitive archaeologist Karenleigh A. Overmann and I believe that one of the most influential and foundational cognitive bases for abstract thinking (it may be framed as a matching-to-sample rule) is the appreciation of numbers (numerosity), which is an ancient primate ability and does not require language or even protolanguage. Being able to generate goals requires the selection attention of the organism, and this ability, too, appears to rely heavily on the PFC (but not exclusively). The standard means of learning, classical and operant conditioning, are available to nearly all life forms; even the relatively simple brain and nervous system of planaria can be classically conditioned. However, these systems of learning usually require many trials, and the learning is relatively fixed. Primates possess both of these learning systems, but they have also evolved a unique system for using single events to form goals based on an object (fruit) or a place (this branch of a tree). As a result, primates can more quickly respond to changing environmental circumstances, thus improving the quality of their choices. Neuroscientists Passingham and Wise (2012) have called this newer PFC-controlled attentional system of learning in primates a "new way of knowing what to do in nonroutine situations" (p. 254). According to Passingham and Wise, when traditional and phylogenetically older learning systems fail, this less automatic, slower system of PFC attentional control takes over, thus allowing for more flexible behavior.

A review of human and animal studies of both normally functioning and brain-damaged individuals reveals that the PFC, particularly the DLPFC, has most of the chief functions of experimental psychologist Alan Baddeley's

(2001, 2012) working memory model's central executive. In Passingham and Wise's view, the chief functions of the DLPFC (and presumably orbito-frontal PFC, to be mentioned subsequently) are the following (plus one that I added):

1. *Integration*—primate brains conjoin information from all sensory and motor systems based on the current problem or context (in Passingham and Wise's vocabulary, "a single event"). The integration of this information enables primates to make better decisions about what resources to obtain, where to go, and where not to go. As visuospatial information requires the place and grid cells of the hippocampus (see Chapters 3 and 8), the DLPFC must rely on some hippocampal functions as well. This function of integration also allows primate brains to categorize the nature of things and, as mentioned earlier, to extract abstract qualities of objects, events, and places to create discreet categories. Abstractive abilities also appear to involve the cerebellum (see Chapter 7).

2. *Interference*—this function of the DLPFC allows primate brains to block stimuli that are irrelevant to completing the task. Notice that this function is an important aspect of attentional control (or they are op-posite faces of the same coin), and it is the same function in a generic working memory model or Baddeley's working memory model. The ir-relevant stimuli may arise from the current external environment of an event or from previous internal memories, and the irrelevant stimuli must be appropriately inhibited from gaining one's attention.

3. *Flexibility*—the ability to shift attention to other tasks and switch back to the original task has long been thought to be a critical cognitive func-tion in the evolution of human brains. Anthropologist Steven Mithen (1996) popularized a synonymous term, *cognitive fluidity*, by which he meant an ability to integrate various forms of knowledge (e.g., so-cial, technical, etc.) so as to construct novel solutions, construct novel artifacts, or solve novel problems. In that regard, flexibility partially overlaps with the integrative function of the PFC; it is important to note that none of these functions should be considered completely inde-pendent or in isolation from one another.

4. *Prospection*—although the word *prospection* literally means to antic-ipate the future and generate future outcomes, *prospective memory* is often more narrowly defined as recalling a memory in the service of a present goal. It might also be colloquially defined as *remembering*

to remember. In the latter definition, prospection would overlap with the executive function of planning and organizing activities to achieve future goals.

5. *Sequences*—being able to recall an appropriate series of subordinate goals in the service of an overarching goal is an important function of the DLPFC. Interestingly, it is usually mentioned when referring to one of the many classical functions of Broca's area, but usually in the lexical context of spoken (and presumably internal) speech. It is now thought that Broca's area does serve this lexical function, and it also appears to have higher-order and supramodal (i.e., transcending or integrating different senses) sequencing functions. The latter ability would obviously be critical to the central executive functions in Baddeley's working memory model. Anecdotally, it is reported that Neurosurgeon William Scoville (see Chapter 8) suspected his sister had a frontal lobe brain tumor because her exquisitely planned Thanksgiving meals suddenly became chaotic, as the ability to cook and prepare a multi-course meal would not only depend on the ability to prepare various foodstuffs but begin each at the appropriate time.

6. *Valuation*—the evaluation of various behavioral possibilities has been shown to be highly dependent on the DLPFC. One of the little-discussed behavioral repercussions of Gage's tragic accident was his inability to assign appropriate monetary value to various goods, including common pebbles. Harlow (1848) wrote: "Does not estimate size or money accurately, though he has memory as perfect as ever. He would not take $1000 for a few pebbles" (p. 392). Of course, Gage's injury probably involved destruction to other prefrontal cortices, so this particular sequela cannot be solely ascribed to the DLPFC, but it is consistent with lesions isolated to the DLPFC. The classic executive function of decision-making has also been shown to be a critical DLPFC function including mundane, moral, and risky decisions.

7. *Inhibition*—although not on Passingham and Wise's list, the ability to inhibit responses appears to underlie all previous six functions. It was one of the most noted psychological problems of Phineas Gage, particularly his capriciousness, irreverence, and profanity (all premorbidly atypical). Aron, Robbins, and Poldrack (2004) have demonstrated in humans and in other animals that DLPFC and orbitofrontal cortices are involved in the ability to inhibit responses, and the right orbitofrontal cortices may be the most critical area, although left hemisphere

areas may also be involved. The DLPFC has been shown to be important in deception (inhibition of true responses) and truth-telling, and
interestingly, the DLPFC shows equivalent activity when preparing to
tell the truth or a lie (e.g., Ito et al., 2012). Inhibition deficiencies (as
measured by transcranial stimulation and EEG) have also been demonstrated in psychopathic offenders compared to typical controls (e.g.,
Hoppenbrouwers et al., 2013). A pseudopsychopathic syndrome was
also coined years ago when patients with frontal head injuries or tumors
became disinhibited, thus the frontal cortices have long been thought
to be responsible for "civilized" behavior (e.g., Goldberg, 2002, 2009).

Orbitofrontal Prefrontal Cortex

About 25 years ago, I received feedback from an anonymous reviewer of
an empirical study that I had conducted on closed-head–injured patients.
I had suggested that executive functions might also be involved in emotional
and social decision-making, besides classical laboratory-defined decision-
making (i.e., without major personal or ego involvement in the task). The
reviewer wrote, "executive functions are COGNITIVE." Research into the
functions of the orbitofrontal prefrontal cortex (OFPFC) has burgeoned over
the past 25 years, and now an interesting opposite reaction has seemed to
have occurred: The OFPFC has been implicated in a wide variety of cognitive
functions and neuropsychiatric disorders, to the extent that it may not be
possible for the OFPFC to be directly involved in so many functions.

One of the most prominent of these newly proposed functions of the
OFPFC is response inhibition. Although a precise determination of the location of Phineas Gage's frontal lobe damage is difficult, the notion that frontal
lobe damage may be associated with inability to inhibit one's responses
has persisted in the literature. Interestingly, this association persists despite seemingly contrary evidence that frontal lobotomies more often result
in apathy and a severe lack of spontaneity. UCLA neuroscientist John Van
Horn and his colleagues (2012) analyzed earlier magnetic resonance images
of Gage's skull and concluded that a significant portion of his frontal white
matter (association cortex) had been damaged, which might have accounted
for Gage's subsequent capricious and uninhibited behaviors. To definitively
decide whether response inhibition is or is not a core function of the OFPFC,

Stalnaker, Cooch, and Schoenbaum (2015) reviewed the OFPFC literature and found many cases where OFPFC damage did not lead to disinhibition, and they concluded that response inhibition was not one of its core functions.

Stalnaker et al. also concluded that the OFPFC was not absolutely required for flexible representations of classically or operantly conditioned responses, although it likely plays some role in conditioning but its specific role has yet to be determined. Another touted function of the OFPFC is its role in signaling emotions, specifically making decisions based on precomputed emotions (i.e., largely unconscious or involuntary emotional reactions), which is the basis for the somatic marker hypothesis (e.g., Bechara & Damasio, 2005). This hypothesis states that emotions and bodily responses like anxiety, fear, or blushing guide decision-making. In this regard, Stalnaker et al. concluded that the OFPFC did play a role in somatic marking, but not so much in using them to trigger decisions but in imagining or guessing emotional responses to future outcomes. Stalnaker and his colleagues endorsed a relatively new critical role for the OFPFC: It may serve as a cognitive map, in conjunction with many other brain areas, that helps to define the relevant cognitive space for a current task. It may be necessary, they surmise, to guide behaviors appropriate to achieve goals. This cognitive map might be used to make predictions about the outcomes for alternative or competing behaviors. It might also enable the quick disregard for old rules in favor of new ones and the ability to determine what aspects might have been at fault when a goal-directed behavior failed. The OFPFC might provide this cognitive-map information to other brain areas such as the hippocampus (see Chapters 3 and 8), but Stalnaker et al. emphasize that the formation of this cognitive map does not rely solely on the OFPFC.

The Exaptation of Fruit-Eating to Tool-Making

As noted earlier, the earliest primates' brains evolved under selection for their ability to recognize ripe fruit in trees, grasp it, and eat it with their forelimbs. As very small nocturnal creatures, it appears they coordinated this activity vocally with conspecifics (same species) in order to compete successfully with other animals who also desired ripe fruit. All folivores (leaf-eating animals) have smaller brain-to-body ratios (called an EQ) than those of frugivores (fruit-eating animals) because leaves are less nutritious and harder to digest. But, of course, carnivores (meat-eaters) and omnivores (those that eat meat and anything else) have an even higher EQ; as has been noted previously, brains are a very expensive metabolic tissue, and only a supplement of meat in a diet would allow the evolution of bigger brains.

It is now thought that the first stone tools (Mode 1; see Fig. 1.2) were mostly sharp flakes made by the australopithecines, beginning about 3.4 million years ago. They were knapped by striking a stone (hammer) on another stone (core). These flakes and cores are referred to as *Oldowan,* for the Olduvai Gorge in Tanzania, where famed archaeologist Louis Leakey found them in the 1930s, and both the flakes and cores were subsequently used as tools. Because brain tissue cannot evolve suddenly and de novo (starting anew), the brain regions that evolved to recognize fruit, grasp it, eat it, and coordinate that activity by vocalizing with other primates had to be exapted to successfully make Oldowan stone tools. In contemporary experimental Oldowan stone tool-making studies, archaeologist Dietrich Stout and Chaminade (2012) and Stout, Hecht, Khreisheh, Bradley, and Chaminade (2015) have shown that the Oldowan stone tool-making technology can be learned by having students' attention drawn to the task at hand while imitating the teacher's movements. Other studies have shown that language, that is, being able to verbally instruct, is not a sine qua non (requirement) for acquiring the Oldowan skills. Stout and his colleagues also proposed that Oldowan tool-making draws heavily on the dorsal stream of circuits from the occipital lobes, anterior and posterior parietal lobes, premotor and supplementary regions, and inferior frontal regions. Remember also that the imitation of a teacher's movements and the comprehension of the intention of those movements might have been aided by mirror neurons (if the basic aspects of the mirror neuron hypothesis are valid), and since those neurons have been identified in macaques, they were likely to have been present in australopithecines.

The knapping of Acheulean (Mode 2; Fig. 1.3) bifacial, symmetrical handaxes, which began to appear about 2 million years ago, obviously made greater cognitive demands than the Oldowan technology (Wynn & Coolidge, 2016), and this is exactly what Stout and his colleagues were able to demonstrate experimentally. In their studies of naïve college student volunteers, they found that the same dorsal stream was activated in the making of handaxes, but there was additional recruitment of the right ventral premotor cortex and the pars triangularis (BA 45; part of Broca's area) in the right hemisphere. The pars triangularis plays an important role in semantic processing and analytical reasoning. Later Acheulean handaxes became thinner and more refined, and they required not only greater core management but also more elaborate hierarchies of reasoning, including the linking of outcomes with subgoals and greater cognitive flexibility. The complexity of these hierarchically linked, multistage strategies for the production of handaxes suggests that the cognitive functions of the prefrontal cortex were under greater selection than ever before. Acheulean tool-making required not only greater attention (effortful cognitive control) but also task-outcome monitoring, greater motor control, and prospective simulation. Thus, as this chapter has attempted to demonstrate, the frontal lobes came under heavy selection pressure beginning with the earliest primates, and this pressure continued through the early history of hominins. It is also possible that the cognitive bases for handaxes were under selective pressure, particularly within the last 1 million to 500,000 years ago. It appears that handaxes during this period were being "overdetermined"—that is, stone knappers began giving their products more attention and greater effort than was necessary to perform their standard functions. Some handaxes were way too large to comfortably fit in one's hand, some had consistent asymmetries, and some were simply more aesthetically pleasing to the eye, either by the stone's colors, shape, or both.

Other Cognitive Advances 2 Million Years Ago

Social learning is an important part of the transmission of cultural traits in human and nonhuman primates. There are certainly some cognitive commonalities between these two animal groups with regard to social learning, but there are also some clear differences. The two groups share the ability to imitate and to share attention with conspecifics. True imitation requires not only reproducing a behavior of a conspecific to achieve a goal but also having an understanding of the intentions of a conspecific for achieving a goal. This is where nonhuman primates may differ from humans, as it is highly debatable whether the former truly "understand" a conspecific's intentions. Simple operant (reward) conditioning can explain an ape's imitation of a conspecific, and Occam's razor (the simplest explanation is preferred) is often *not* invoked by primatologists. Also, as noted previously, the mirror neuron system may provide a neural mechanism for imitation with regard to goals, but caution is urged when attempting to determine whether it is also a mechanism for understanding the *intentions* of a conspecific. Primatologists and others often differentiate between imitation and emulation, where in the latter process a primate attains the same goal as a conspecific, but achieves that goal through any means, rather than through the specific behavior of the conspecific, to achieve the same goal. Regardless of whether nonhuman primates are capable of true imitation or have a greater predilection to emulate in order to achieve a goal, clear differences between the two groups persist. The same may be true of shared or joint attention. Humans appear to differ qualitatively and quantitatively from nonhuman primates. Human infants, but not adult apes, readily and frequently share attention with adults, and they often do so even for the purpose of just sharing attention. Only in rare cases like autism spectrum disorder are humans deficient in this ability.

As noted in Chapter 1, about 2 million years ago, *Homo erectus* began congregating in much larger groups than the australopithecines and habilines. Thus began an opportunity for group members to learn socially from more experienced members. It served as a means of cultural transmission across generations through which skills of all types could accumulate. However, as cognitive archaeologist Ceri Shipton (2010) has argued, these primary means of social learning, imitation and shared attention, may not have been directly affected by group size (although, in my opinion, they must have been at least indirectly affected by group size). He hypothesized that the underlying neural mechanisms must have been in place, which is the central thesis of this book and this chapter. He also claimed that reliable and high-fidelity (i.e., true to the original intention) social learning allows a cultural ratchet effect such that innovative technologies do not have to be reinvented with each generation and provide for increasing cultural complexity that all members may benefit from. As noted in Chapter 2, neural reuse may be one of the most important organizing principles of the brain. This confluence of many selection pressures—the transition from tree life to full-time ground life, vastly increased group size, the advent of fire, and demands for a social brain—may have acted in a concerted way (i.e., concerted evolution) on not only individual brains regions with particular functions (i.e., mosaic evolution) but also on networks of brain regions such as the frontal-parietal network and others. One final issue, in light of this chapter, is whether the exaptation of frontal-parietal networks for tool use co-instigated their exaptation for language or protolanguage, but this will be addressed in later chapters.

Summary

1. The expansion of brains in mammals, particularly the cortex, increased behavioral flexibility.
2. Mammalian brains have been typified by mosaic evolution (natural selection for particular brain structures and neurons independent of other regions) and concerted evolution (natural selection for brain structures and networks with complimentary functions).
3. Two important influences on modern human brains have been the evolution of mammalian brains and primate brains; the latter has had the most profound influence.
4. The prefrontal cortex (PFC) is one of the major exaptations of the human brain, where executive functions primarily reside.
5. The major executive functions of the PFC are decision-making, forming plans and goals, organizing, devising strategies to attain goals, inhibition, and the monitoring of effective performance.
6. The frontal lobes of the earliest primates were under selective pressure to identify and eat fruits with their forelimbs. The brains of hominins may have exapted these same regions for object manipulation and tool-making.

5

The Parietal Lobes

If brains are for *moving*, then it would have been critically important for inchoate organisms to know where they were moving from and to. It appears
that the original adaptation of the parietal lobes was the gathering and
integration of sensory data to aid the motor cortex in these movements
(visual- and sensory-motor coordination). It is not a mere coincidence that
the premotor and motor cortex in the posterior portion of the frontal lobes
is directly adjacent to the somatosensory cortex in the anterior portion of
the parietal lobes. The relative importance of sensory information can be
gleaned from the following cortical sensory homunculus ("little person")
based on a drawing by neurosurgeon Wilder Penfield (Fig. 5.1). This distorted body image shows that hands, lips, tongue, and face have more cortical regions devoted to them than to the trunk or toes. It may also reflect
the relative importance of the original adaptations in human evolution, thus
hands and face undoubtedly played a more significant role in human evolution than trunks or toes. Damage to the parietal area profoundly affects
one's body image or perception of one's body. For example, damage to the
parietal region controlling one's contralateral fingers may result in a type
of *agnosia* (literally, "a lack of knowledge"), which is a failure to be able to
distinguish among one's fingers (*finger agnosia*) or between one's body parts
(*autotopagnosia*).

The lateral superior portion of the parietal lobe (BA 7) is called the *superior parietal lobule* (SPL), and it is well defined inferiorly by the intraparietal
sulcus, which runs horizontally from the anterior SPL and then at an angle
(obliquely) to the posterior portion of the SPL (Fig. 5.1). The medial portion of BA 7 is called the *precuneus,* and little was known of its functions
until more recent neuroimaging techniques were available. Inferior to
the IPS is the inferior parietal lobule (IPL). Its two major regions are the
supramarginal gyrus (SMG; BA 40) and the angular gyrus (AG; BA 39). See
Figs. 3.3 and 5.2.

Evolutionary Neuropsychology. Frederick L. Coolidge, Oxford University Press (2020).
© Oxford University Press.
DOI: 10.1093/oso/9780190940942.001.0001

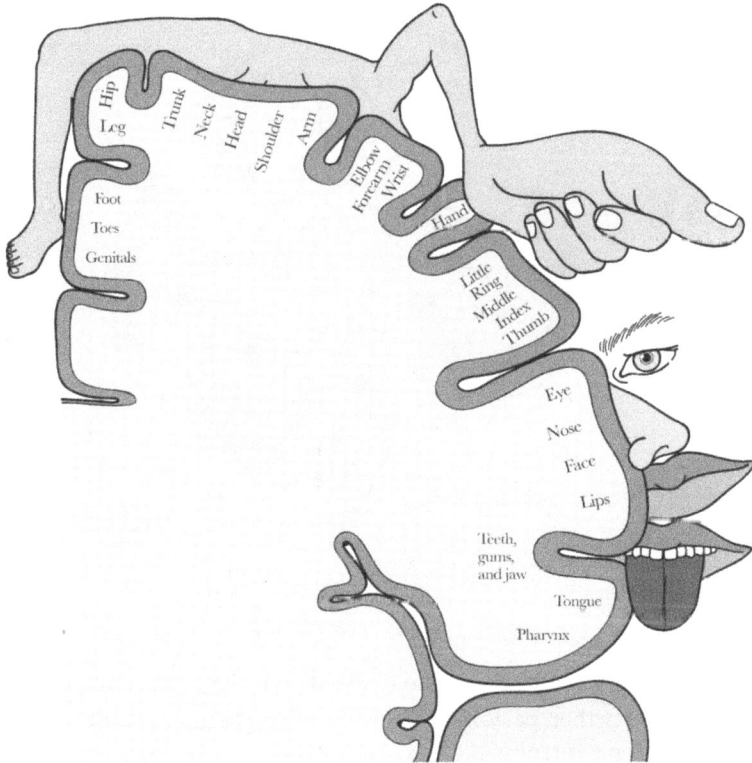

Fig. 5.1 A cortical sensory homunculus.
Source: Holly R. Fischer, MFA.

Brain Evolution in the Past 250,000 Years

It was long suspected that *the* major difference between modern human brains and earlier hominins was an expansion of the frontal lobes in modern *Homo sapiens*. It is now suspected that there was no expansion of frontal lobes relative to other lobes that characterizes modern human brains but rather an expansion of the parietal lobes, particularly between modern *Homo sapiens* and our extinct cousins, the Neandertals. It is important to note, however, that expansion may only be part of the story, as frontal lobes in modern *Homo sapiens* may have evolved a different cytoarchitecture than that of other hominins and co-evolved with other brains regions in a different

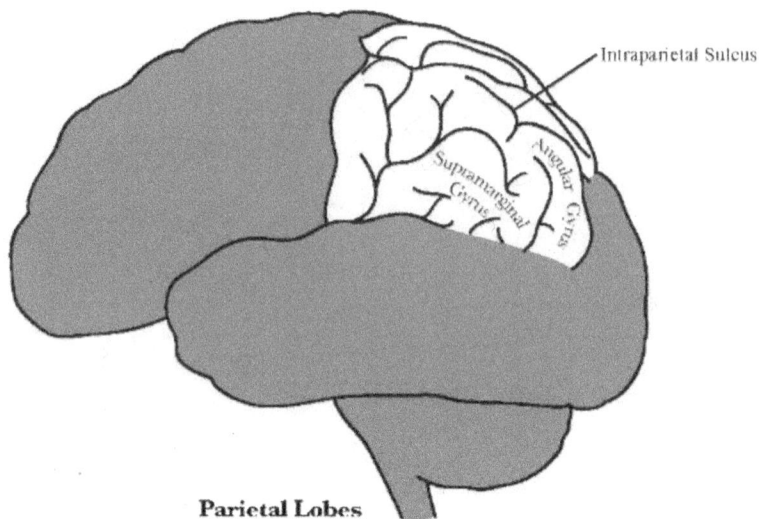

Fig. 5.2 Parietal lobes.
Source: Holly R. Fischer, MFA.

manner than any other hominin. So to dwell on brain expansion per se may be misleading. Further, paleoneurologist Emiliano Bruner's work since 2004 (e.g., Bruner, 2004, 2010; Bruner, Amano, Pereira-Pedro, & Ogihara, 2018; Bruner & Iriki, 2016) has demonstrated a clear expansion of the parietal lobes in modern *Homo sapiens*, particularly the SPL and precuneus, compared to these regions in *Homo erectus* and Neandertals (despite the latter having a larger brain than that of modern *Homo sapiens*). Bruner and others (Neubauer, Hublin, & Gunz, 2018) now claim that this parietal lobe expansion may have occurred relatively recently in the evolution of modern *Homo sapiens,* within the last 100,000 years.

Other Parietal Regions

Besides the classic somatosensory cortex of the anterior portion of the parietal lobe (BA 1, 2, and 3) and the somatosensory association cortex (BA 5), many other regions have been exapted for important higher cognitive functions. Most of these regions retain their phenotypic functions (a phenotype refers to either observable brain structures or behavior) based on their original adaptations, but the extent to which they retain these functions while reorganizing neuronally to perform new functions is not fully understood. These other regions include superior regions of the parietal lobe, the IPS and precuneus (BA 7), and inferior regions such as the SMG (BA 40), AG (BA 39), and some inferior adjacent regions like the retrosplenial cortex. If the classic function of the parietal lobes was "knowing" where a primate was in a tree and directing its movements elsewhere, then it is provocative to think that the exaptations of the parietal lobes help humans to know who they are, where they are, and where they are going, both in a physical and in a metaphorical sense (e.g., Lou et al., 2004). For example, much has been made of the parietal lobes as a critical part of the default mode network of the brain; that is, if a person is placed in an fMRI scanner and asked to relax before a noninvasive experiment, regions of the parietal lobes become active. It is as if the parietal lobes become active when a person ponders what is going on, what is about to happen, and drifts from thought to thought. I will return to the default mode network and discuss it in greater detail later in this chapter.

Intraparietal Sulcus

First, it is important to explain why the sulci (plural of sulcus and pronounced *sul-sigh, sul-key,* or *sul-kus*) are mentioned so frequently in brain studies. Formally, the cortex is only the top six layers of the surface of the brain, which is about the thickness of an orange rind. If a brain has a relatively smooth surface (without many gyri or sulci), it is called *lissencephalic*. Some animals, like rats, cats, and squirrel monkeys, have relatively lissencephalic brains, and their behavior is generally simpler than that of animals who have highly convoluted brains, like dolphins and chimpanzees. There is also a rare brain disorder in humans in which their brains are lissencephalic. Sadly, those humans with the disorder are usually profoundly disabled, often mute, and have a very short lifespan. Thus, the convolutions of the brain form the sulci, and the surface of the sulci create a greater overall surface area than a lissencephalic brain. It is estimated that about 40 to 50% of the cortex is contained within the sulci, and the parietal lobes are particularly convoluted, which may be an important evolutionary marker in terms of higher cognitive processing.

The IPS is present in humans and monkeys, although in humans it appears to have expanded beyond that expected from body size. The anterior portion of the IPS (horizontal IPS, or hIPS) has been shown to respond to numbers.

As noted in Chapter 3, number appreciation is called *numerosity*, which is a capability shared by humans and many animals. There are neurons in the IPS that are attuned to the two core numerosity processes, subitization and the ability to distinguish between smaller and larger sets. Also, as noted previously, we (Coolidge & Overmann, 2012) theorized that numerosity serves as the cognitive basis for abstraction, is independent of language, and was important evolutionarily in predator avoidance and food gathering. The IPS may also be responsible for determining the serial order of numbers or letters. For example, the left and right hemisphere IPS is active when humans are asked to determine whether 3 comes before 7 or if F comes before C. It has also been shown that different specific sets of neurons in the IPS are active depending on whether the task involves numbers or letters (Zorzi, Di Bono, & Fias, 2011), although tasks that require ordering either numbers or letters may involve a domain-general comparison process in only the left IPS (Attout, Fias, Salmon, & Majerus, 2014). It has been argued that this "sense" of numbers resembles a primary sense along with vision, audition, touch, taste, and smell (Coolidge, 2018; Harvey, Klein, Petridou, & Dumoulin, 2013). Harvey and his colleagues also found a topographic (maplike) neuronal specialization for numbers in the parietal region, and they argued that the processes involved in numerosity are abstract in their nature, which serves as evidence that numerosity is a "higher cognitive function" (p. 1126). Recently, Kutter and colleagues (2018) found evidence that single neurons and groups of neurons may respond to symbolic numbers (e.g., 7) and nonsymbolic numbers (e.g., + + +).

Supramarginal Gyrus

As noted in Chapter 2, neuroimaging studies of the SMG (BA 40) have shown it to be important in the temporary storage of sounds (phonological storage) and in the linkage of sounds to meaning. Crowder and Morton (1969) called this ability *precategorical acoustic storage* (PAS). It is precategorical in the sense that the sounds can be temporarily stored without regard to their meaning, be they nonsense syllables or words from a foreign language. They likened it to a short-term verbal memory store of a few seconds, apparently long enough to apply meaning to the sounds or determine that the sounds are unfamiliar. Conceptually, PAS is much like Baddeley's working memory component, the phonological loop (although he objects to the comparison).

The sounds are held in storage long enough to match them to a meaning or to discover that there is no match (such a nonsense syllable or novel word). Phonological storage is valuable because it provides a means of repeating the sound, either vocally or subvocally (by means of inner speech), in order to memorize it. Baddeley called this component of phonological storage the *articulatory loop*. It has also been proposed that the SMG is part of a network involved in inner speech, along with the superior temporal lobes and other brain regions (including the cerebellum [see Chapter 7]). The SMG is also engaged in verbal working memory tasks in which continuous verbal information must be temporarily maintained and attended to despite interference (e.g., Deschamps, Baum, & Gracco, 2014).

Parts of the SMG may also have other unique roles. The posterior portion of the SMG (SMGp) helps to detect whether a person is upright or not, partially based on the position of one's head, orientation of the eyes, and input from the retinas. Kheramand, Lasker, and Zee (2013), in an empirical study of humans, concluded that the SMGp integrates different sensory input, creating a "common spatial reference" (p. 6). The right SMG (rSMG) may play an important role in emotional processing. Silani, Lamm, Ruff, and Singer (2013) found that the rSMG provides an additional cognitive function in emotional processing. They claimed that although emotional understanding is subserved by shared neural networks, the rSMG may play a unique additional role by helping to avoid emotionally biased social judgments. They reasoned that the ability to be empathic is largely a function of having experienced that emotional experience. Note that this simple self-projection mechanism may lead to an egocentric bias. They proposed that when one is aware that one's self-projection may potentially be biased, the rSMG helps to detect these situations and provide input intended to correct them.

Orban and Caruana (2014) found that the left anterior portion of the SMG (aSMG) is active when humans observe other humans using tools, but not when macaques see their conspecifics using tools. These authors noted that the aSMG is part of an object-manipulation network in which it receives dorsal and ventral streams of visual information, as well as non-semantic information about to-be-manipulated objects from the posterior parietal lobes and the anterior intraparietal sulcus. Orban and Caruana theorized that modern *Homo sapiens* may have evolved two distinct parietal systems: (1) an older biological brain network for grasping and manipulating objects, and (2) an emergent (more recent) parietal system, with converging sources of information including semantic information, specifically devoted to tool use.

Further, they claimed that it was the latter system that aided humans in understanding causal relationships between tool use and specific goals. In this manner, they proposed that tool use may "have triggered the development of technical reasoning in the left IPL, which in turn may have favored a development of language in the left hemisphere" (p. 9).

Phonological Loop

As noted earlier, F. Aboitiz, S. Aboitiz, and Garcia (2010) have proposed that the SMG, in conjunction with temporal, inferior parietal, and frontal areas (BA 22, BA 39, BA 44, BA 45), forms a specialized auditory–vocal circuit that was a fundamental element in the evolution of language. They view the phonological loop as a "key innovation" in the evolution of language. As has been asserted thus far, the emergence and evolution of primates had one of the most profound influences on modern human brains and language, launching humans into a "cognitive niche" (e.g., DeVore & Tooby, 1987). Because brains are composed of such metabolically expensive tissue, they could not have expanded based on the meager nutrition provided by leaves and grass. Primates competed with other animals for far more nutritious fruits. However, because of these early primates small size, they competed more successfully through social vocalizations with their conspecifics. The ability to store these sounds and apply "meaning" became critical to their success, allowing the expansion of their brains relative to their body size (increasing their EQs, although there are many other factors accounting for general intelligence in addition to EQ). From the time of these early primates through the ancestors of modern humans, the development of the phonological loop may have enabled increasingly longer and more complex utterances and the application of meanings to them. Given that extant vervet and putt-nosed monkeys teach specific, meaningful sounds to their young, it is not hard to imagine that early and much bigger-brained hominins had a large array of sounds resulting in a kind of protolanguage even in the earliest of australopithecines and habilines. Certainly, a kind of protolanguage, vocal communications and/or hand gestures, was well in place at the emergence of *Homo erectus*. Aboitiz et al. (2010) have proposed that this continued use and exaptation of the phonological loop promoted social bonding and allowed the coordination of various kinds of activities, ultimately resulting in increasingly complex communications. Further, it may have enabled the creation of relationships among strings of sounds, or phrases, which may have fostered thoughts within thoughts. The latter phenomenon is called *recursion*; it is said to be the hallmark of modern language (see Coolidge, Overmann, & Wynn, 2010, for an overview of recursion). The evolution of modern language will be discussed in greater detail in Chapter 6.

Angular Gyrus

The AG (BA 39) sits posteriorly but adjacent to the SMG and superior to the posterior portion of the superior temporal gyrus. One of its first functions was discovered in 1891by French neurologist Joseph Jules Dejerine, who found that lesions to the left AG (and SMG) resulted in an inability to read (called *alexia*). More recently, when the right AG was electrically stimulated it produced out-of-body experiences. Californian neurologist V. S. Ramachandran and his colleagues have suggested that the AG was part of a neural network for the production and understanding of metaphors, although that finding is still controversial. There is, however, substantial evidence that the AG plays a role in mathematics, and damage to the AG has been shown to impair arithmetic operations (called *acalculia* or *dyscalculia*). Contemporary English neurologist Mohamed Seghier (2013) reviewed the neuroimaging literature on the AG and found it to be involved in at least 15 major cognitive functions and that it may interact with at least 15 other regions of the brain. Further, his review established that the AG may be parcellated (divided) into many distinct regions based on cytoarchitecture, connectivity to other brain areas, and functions. Among its many functions, the AG seems to be a cross-modal hub that integrates sensory information (e.g., visual, auditory, tactile) and gives appropriate meanings to events, manipulates visuospatial representations of the environment, attends to salient aspects of the environment, and solves various problems (mathematical, reasoning, social-cognitive). Cognitive neuroscientist Antonio Damasio (1989) has labeled the AG and related areas *convergence zones*, where the zones' basic nature is multimodal and serves as a gateway for accessing different kinds of sensory information, integrating that information, and creating abstractions of events into concepts that allow for further processing. Finally, and perhaps not surprisingly, as noted in Chapter 2, the AG has been implicated in being part of the "default mode network" of the brain that becomes active when humans are resting and apparently involved in their own thoughts.

Precuneus

As noted in Chapter 2, the medial and superior portions of the parietal lobe (BA 7) have received much recent attention, especially the medial portion known as the precuneus. Because of its location, tucked into the middle and sides of the superior portion of the parietal lobes, it was relatively difficult to study until the advent of more sophisticated neurophysiological studies. Fletcher et al. (1995) conducted one of the first neuroimaging studies to implicate the precuneus in the recall of visual memories but not in semantic memory recall. This recall of visual memories is called *episodic memory,* as coined by experimental psychologist Endel Tulving (1972, 2002). He noted that this ability to not only recall personal memories but also manipulate them and simulate future scenarios was critical in the evolution of modern thinking and problem-solving. Episodic memories may be thought of as memories for episodes or scenes in one's life. They generally have time, place, and meaning characteristics. The latter is important in memorization because emotionally salient experiences are learned much more quickly and over longer periods of time than less salient experiences. Also, it is important to note that animals from foxes to ravens undoubtedly have exhibited signs of episodic memory (Allen & Fortin, 2013), so the same seems to be true of them. Where food is cached has much greater meaning (evolutionary survival) than more mundane experiences.

Tulving (1972) has investigated the differences between episodic memory and semantic memory (the latter being a memory for words and meanings). In 2002, he elaborated on the concept of episodic memory by introducing *autonoetic thinking,* an awareness that our memories are relative and that we may go backward and forward in time among them. He reasoned that autonoetic thinking might have been critical to human evolutionary survival, as one could imagine future options and choosing among them without the dangers inherent in trial and error.

If animals can experience at least simple episodic memory, then they must have a precuneus somewhat similar to that in humans (e.g., Margulies et al., 2009). Nonhuman primates and other animals also have connectivity-based subdivisions within the precuneus (just like humans), with anterior, medial, and posterior regions of the precuneus connected to various other brain regions. There is also evidence that the precuneus in both humans and monkeys is part of the default mode network of the brain. Cavanna and Trimble (2006) were able to demonstrate various functions of different regions in the precuneus in healthy adult humans, such as the anterior portion subserving self-centered mental imagery tasks, and the posterior portion being more heavily involved in episodic memory retrieval. They also summarized the results of neuroimaging studies that established the role of the precuneus in visuospatial imaging (critical to and synonymous with Baddeley's visuospatial sketchpad), episodic memory retrieval, and tasks in which one takes an egocentric perception of one's self (self-consciousness), others, and the world and perceives agency in those relationships (i.e., intentionality). Lou et al. (2004) demonstrated through neuroimaging that medial portions of superior parietal lobe (i.e., the precuneus), along with medial portions of the prefrontal cortices, are critical to self-representation and other-representation. In an interesting experimental paradigm, healthy human subjects (from Denmark) were asked to think of themselves, their best friends, and the Queen of Denmark. The investigators found differential activation across the thinking tasks in varying regions of the precuneus, and they found some left versus right hemisphere differences as well, with a greater role for right precuneal activation when thinking of one's physical self and self-representation overall.

The Precuneus and Neandertals

As noted earlier in this chapter, Bruner and Iriki (2016) claim there was an expansion of the parietal lobes in modern *Homo sapiens*, particularly in the region of the precuneus, compared to parietal lobes in Neandertals. *Homo sapiens* and Neandertals shared a common ancestor, *Homo heidelbergensis*, but about 580,000 years ago the two lineages diverged. In 2014, Bruner and Lozano noted that 100% of the teeth of both *Homo heidelbergensis* and Neandertals showed striations, indicating heavy use, while only 46% of the teeth of penecontemporaneous (living at almost the same time) *Homo sapiens* showed such wear patterns. They hypothesized that Neandertals may have had a mismatch between their cultural practices (cleaning hides, processing food stuffs, etc.) and their neural substrates, particularly the superior parietal regions responsible for visuospatial integration. They hypothesized that this mismatch may have required Neandertals to use their teeth as a kind of "third hand" to process hides and perform other duties. They also reasoned that *Homo sapiens,* with expanded parietal lobes, managed to solve these "household" duties without the constant recruitment of their teeth. If Bruner and Lozano are correct, then it might also be reasonable to assume, based on the earlier reviews in this chapter of the functions of superior and medial portions of parietal lobes, that Neandertals might have had a different sense of self-representation and other-representation than that of their contemporary cousins *Homo sapiens*. If that supposition is correct, it might explain the apparent predilection for cannibalism that Neandertals showed over a period of about 80,000 years (120,000 years ago to about 40,000 years ago). Agustí and Rubio-Campillo (2016) have proposed that Neandertals might have been responsible for their own extinction because of a predilection for cannibalism that invading *Homo sapiens* did not share. It has also been noted (e.g., Rougier et al., 2016) that signs of Neandertal cannibalism persist even when other animal meat appears to have been plentiful. The latter evidence led Dieringer and Coolidge (2018) to hypothesize that Neandertal cannibalism might have been a gustatory experience ("this meat tastes good") as well a pragmatic practice of keeping other bands of Neandertals out of one's area. While most modern folks do not engage in cannibalism—most people are repelled and horribly offended by that practice—there are Neandertal cannibalism sites where baby and juvenile Neandertals (along with adults) were butchered and eaten. Thus, there is the possibility that this practice would be less offensive if Neandertals had possessed a different sense of self-representation and a different sense of other-representation based on neural differences in the precuneus. As will be discussed in Chapter 8, Neandertals may also have had an altered sense of smell, which may have made them less disgusted by the smell of burning human flesh.

Retrosplenial Cortex

The cingulate cortex (or cingulate gyrus) is inferior to the prefrontal cortex (PFC) and runs from anterior to posterior in the medial part of the inferior cortex. The cingulate cortex covers the corpus callosum (see Chapter 3, Fig. 3.6). The posterior portion of the cingulate cortex (BA 29 and 30) is called the retrosplenial cortex (RSC). It is called the RSC because it is posteriorly adjacent to the splenium, which is the most posterior portion of the corpus callosum, and the splenium is the thickest part of the corpus callosum. The RSC is part of a network with critical connections to the hippocampus, parahippocampus (surrounds the hippocampus), frontal lobes, and thalamus. The RSC's chief function is the creation and storage of spatial information for spatial navigation. Damage to the RSC results in verbal and visual memory problems, including those involved in the retrieval of recent episodic and autobiographical memories (but not old autobiographical memories). One of its most critical purposes with evolutionary implications appears to be a translational responsibility for processing egocentric spatial viewpoints (a view from one's own perspective, which engages the precuneus as well) and allocentric spatial viewpoints (a view from other perspectives) (e.g., Vann, Aggleton, & Maguire, 2009; Zaehle et al., 2007). Zaehle et al. proposed that both egocentric and allocentric frames of reference are processed by a hierarchical brain network involving the hippocampus, thalamus, precuneus, and frontal and parietal cortices, but the egocentric frame of reference only activates part of the larger network.

Auger, Zeidman, and Maguire (2015) have found that the RSC is responsible for creating, storing, and navigating new environments. As the RSC is common to humans, the great apes, monkeys, and even rodents, it appears that forming a dependable mental map of a new environment was an important part of effective locomotion throughout evolution. However, the translational function of the RSC in humans between egocentric and allocentric frames of reference makes that exaptation of this multicomponent brain network especially important, as *Homo erectus* repeatedly left Africa beginning at least 1.4 million years ago. Further, imagine the evolutionary value of being able to transform an egocentric perspective of the environment into an allocentric one, where other viewpoints can be explored mentally without the dangers of actual trial and error. Further, imagine the value of recalling a location from an episodic memory and being able to translate it into an egocentric viewpoint so that one can act on it from one's own perspective. According to Vann et al. (2009), this is precisely the major function of the RSC—"acting as a short-term buffer for the representations as they are being translated" (p. 797).

Constructive Memory

Perhaps one of the major exaptations of the human brain has been the neural reorganization of the parietal lobes to reflect on one's self and others, which allows for a kind of critical self-awareness, that is, an ability to take note of one's own behavior and to correct it. It has been argued that the essence of the disease schizophrenia, regardless of its specific emotional and cognitive difficulties, is the loss of critical self-awareness. This self-reflective process includes the ability to be aware of prior experiences, both mistakes and successes, and to use them to construct and guide future behaviors. As noted previously, this is the essence of the evolutionary success afforded by an episodic memory system. Many scientists have pondered the prominent role of episodic memory in human evolution. Tulving (2002) noted the sole exception to time's irreversibility appears to be one's episodic memory, the ability to travel back in time in one's mind and even alter previous outcomes. He further proposed that this kind of mental time travel was phenomenologically different from human's actual experiences, and their awareness of this difference is also an aspect of his concept of autonoesis. He also emphasized that episodic memory is an ability to not only call up the past but also imagine future scenarios.

Cognitive psychologists Daniel Schacter and Donna Rose Addis (2007) have also noted that episodic memory did not evolve to recall exactly what happened in one's past, as anyone with a sibling has discovered when discussing aspects of old memories ("It was red!" "No, it was green!"). Schacter and Addis argue that episodic memory is fundamentally constructive by recalling bits and pieces of a past event and reconstructing it. Of course, such representations are fraught with errors, as what one wished would have happened provides that memory with errors and illusions. They also propose that such an essentially fragile system would be useless unless it could be used to construct future scenarios based on past experiences. They have labeled this hypothesis *constructive episodic simulation*, and they note that the neural bases for it are similar to those regions for recalling the past and imagining the future (primarily the hippocampus and parietal and frontal regions). However, they posit some distinctions, as future simulations call on additional brain regions, such as the right PFC and left ventral PFC. One of Schacter and Addis' (2007) most provocative aspects of their hypothesis is that the episodic memory system did not evolve to simply reminisce but to simulate the future. To do so, a system is required that "flexibly extracts and re-combines elements of previous experiences" (p. 1375). Further, they argue that the memory distortions and illusions that co-occur with reminiscences are a byproduct of this flexible and recombinatory process.

The Default Mode Network

The default mode network (DMN) of the brain, as discussed earlier, appears to be active when a person is resting and not engaged in goal-directed thoughts or external activities. The mental activities that activate the DMN include daydreaming, mind wandering, meditating, imagining personal future events, moral judgments, theory of mind, and spontaneous thinking. When a task requires external goal-directed attention, activity in the DMN is suppressed so these two states appear to have an internally competitive relationship; that is, activity in one suppresses activity in the other to a large extent (e.g., Spreng, Stevens, Chamberlain, Gilmore, & Schacter, 2010). Overall, neuroimaging studies strongly suggest that the basic neural substrate for the DMN is a frontoparietal network, but also includes the posterior cingulate cortex, the precuneus, the RSC, the left and right medial temporal lobes, and the medial PFC. Interestingly, monkeys, when not actively involved in their own environments, also have a DMN, in similar brain regions as humans' DMN (Mantini et al., 2011). DMN neuroimaging studies and neuroimaging studies in general are not without controversy, as an fMRI study of a dead salmon showed brain activity (Bennett, Baird, Miller, & Wolford, 2010; note: they later ate the salmon), and fMRI studies have shown unacceptably high false-positive rates (e.g., Eklund, Nichols, & Knutsson, 2016), so additional research of the DMN is certainly warranted. Chapter 9 will also address the DMN's role in sleep.

The Evolution of Psychopathology

The power of an episodic memory system coupled with self-reflection is not without costs. Although humans may be unique in this regard (future simulation and self-reflection), and it may have ultimately driven other species (wooly mammoths) and other human types (Neandertals) to extinction, undoubtedly there are psychological costs. It has long been proposed that anxiety and depression may result from remembering one's past, and it can result from imagining one's future, including thoughts of death (e.g., Perls, 1969). Anxiety and depression have been called the common colds of psychopathology. Certainly, there is evolutionary value in some anxiety, as it may motivate people to work harder, study harder, or put more effort in an activity. Depression, too, may have had evolutionary value, as it may serve as a kind of "time-out," when one temporarily rests, self-reflects, and is taken out of the "rat race" or social competition. However, just as Schacter and his colleagues proposed that memory distortions and illusions are symptoms of an episodic memory system whose primary purpose is to simulate the future

(Addis, Wong, & Schacter, 2007), so, too, may anxiety and depression be psychopathological byproducts of this same system. Their primary evolutionary roles may have included motivation and self-reflection, but that may have been the inadvertent costs of a flexible and recombinatory episodic memory system. Thus, mental time travel and future simulations may have been critical in the ultimate survival of the extant human species and, perhaps, in the extinction of other species, but it was not without psychological costs. It is no wonder that virtually all humans will suffer from anxiety and/or depression in their lifetimes. It is the result of a powerful and highly adaptive memory system, but its ultimate cost is often early death, as people get ulcers, commit suicide, and wage wars, often against completely imagined threats. Psychiatrist Tim Crow (2000) has postulated that schizophrenia may be the cost of *Homo sapiens* possessing language. It is interesting to ponder whether the episodic memory system, particularly autobiographical memory, gave rise to a plethora of psychopathologies. I have labeled these psychological problems in the ancestral environment, *paleopsychopathologies,* and their evolutionary origins will be discussed in greater detail in Chapter 10.

Summary

1. The original adaptation of the parietal lobes was the gathering and integration of sensory data to aid the motor cortex in appropriate movements.

2. Damage to the parietal regions often produces a type of agnosia, in which patients misidentify their fingers or body parts.

3. The parietal lobes may have expanded in *Homo sapiens* compared with their size in Neandertals, and this expansion may have occurred within the last 100,000 years.

4. The intraparietal sulcus (IPS) has among its suspected functions numerosity, which is an appreciation of numbers. The IPS may have single neurons or bundles of neurons dedicated to be responsive to symbolic and nonsymbolic numbers.

5. The supramarginal gyrus plays a major role in inner speech, phonological storage, and emotional processing.

6. The angular gyrus plays a major role in mathematical operations and may serve an important role in as many as 15 other cognitions functions.

7. The precuneus is a critical region for episodic memory, Baddeley's visuospatial sketchpad, and self- and other-representations.

8. The posterior portion of the cingulate cortex is the retrosplenial cortex, which translates egocentric spatial and allocentric spatial viewpoints. This translational responsibility was critical in the evolution of hominin navigation.

9. The constructive simulation hypothesis proposes that the episodic memory system may have evolved not for perfect scenario recall but for the ability to manipulate past events for future success.

10. The parietal lobes are an important part of the default mode network of the brain. The default mode network is active when a human or nonhuman primate is resting and not engaged in a specific mental activity.

11. The episodic memory system's, whose primary purpose may have been simulating the future, may have also given rise to various psychopathologies such as anxiety and depression.

6

The Temporal Lobes

As nonhuman primates can communicate but do not possess anything like human language, the exaptation of the temporal lobes for modern, symbolic, recursive language must be considered at least a minor miracle. At the outset of this chapter, however, it must be noted that discussion of language, even its basic definition and evolutionary trajectory, inspires vehement arguments among scholars. The temporal lobes, as noted in Chapter 2, consists of BA 20, 21, 22, 36, 37, 38, 41, and 42 (see Figs. 3.3 and 3.7) and are demarcated from the frontal lobes by the lateral (Sylvian) fissure. The temporal lobes lie inferior to the parietal lobes and anterior to the occipital lobes. Its neurons have been exapted to process language and apply meaning to sounds. Thus, the temporal lobes play the major role in thinking, inner and external speech, and memory, in conjunction with other brain networks and regions like the hippocampus. As noted previously, the three gyri (superior, medial, and inferior) of the temporal lobes are anatomically prominent, and the three gyri are perhaps even more distinguishable in chimpanzees, thus, almost ensuring that *Australopithecus afarensis* (Lucy) had temporal lobes similar, at least in gross shape, to modern humans (Fig. 6.1). That the temporal lobes of the australopithecines and *Homo erectus* processed the meaning of sounds and were highly active when communicating is also assured, although whether this neural substrate was already exapted to process language or even a kind of protolanguage at that time (i.e., 2 to 3 million years ago) is debatable. It is known, based on studies of extant monkeys and chimpanzees, that even specific structures within the temporal lobes, such as the fusiform gyrus, have similar functions as in humans, which is the recognition of objects and faces.

Broca and Wernicke's Areas

The French physician Jean-Baptiste Bouillaud (1796–1881) was an early proponent of Gall's phrenology, as Bouillaud was convinced that cerebral functions, particularly spoken speech, could be localized in the brain. In

Evolutionary Neuropsychology. Frederick L. Coolidge, Oxford University Press (2020).
© Oxford University Press.
DOI: 10.1093/oso/9780190940942.001.0001

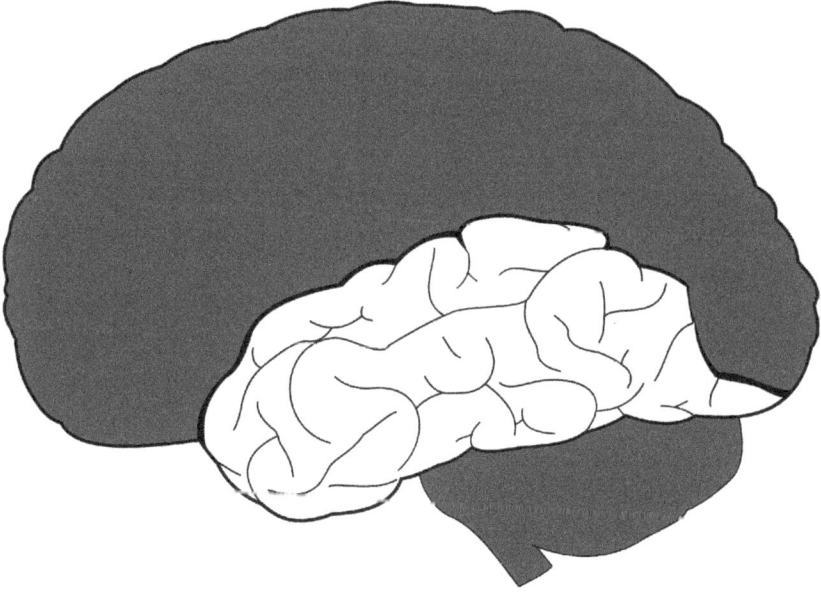

Fig. 6.1 Temporal lobes.
Source: Holly R. Fischer, MFA.

1825, he maintained that this ability could be localized to the anterior portion of the brain, but he broke with Gall, as Bouillaud thought measuring
head bumps was not a viable means localizing the brain's functions. Despite
Bouillard basing his opinion on hundreds of cases of various speech pathologies, the scientific blowback against phrenology remained so strong at that
time that any claims for the localization of brain functions were ignored or
dismissed. In fact, by 1840, others had noted many cases where aphasia (i.e.,
a speech of language disorder) did not result from damage to the anterior
portion of the brain, and so Bouillaud's brave claims went largely ignored.
In 1861, a patient named "Tan" was brought to the attention of French neurosurgeon Paul Broca (1824–1880). Tan was nearly mute and could only
produce the sound "tan" and some obscenities. Only 6 days later, the patient
died, and Broca's detailed autopsy revealed a severe lesion in the inferior portion of the frontal lobes. Broca's report specifically noted that the patient's
articulate speech was disrupted by the lesion. In part, based on the highly
specific details of Tan by Broca and because of Broca's apparently august reputation within many scientific communities, including founder and secre-
tary of the Society of Anthropology (in Paris), he managed to sway many

scientists to accept the idea that some cognitive functions could be localized in the brain. He called the loss of articulate speech due to damage in the left hemisphere *aphemia*, which was subsequently renamed *Broca's aphasia*. Interestingly, Broca was one of the first to examine some skeletons and artifacts found in a cave (Cro-Magnon) near Les Eyzies, France. Based on the size of their skulls and the volume of their endocraniums (inside volume of their skulls), he concluded they were a completely different race, and he hypothesized that they were "intelligent" with "fine cerebral organization" and possessed enough idle time to "cultivate the arts" (Schiller, 1992, p. 156). The skeletons were, of course, *Homo sapiens*, dating to about 32,000 years ago (and are often called *Cro-Magnon* people). Their strikingly modern culture was called the *Aurignacian*, which included elaborate personal ornaments, depictive cave paintings, creative figurines, and highly ritualized burials.

In 1869, an English neurologist, Henry Bastian, presciently proposed that thinking was done in words and that auditory brain centers first received these "sound impressions." He also proposed that there might be different types of aphasia depending on what brain areas were damaged. He described the difference between "word deafness" and "word blindness." Unfortunately, he did not publish this major work for 19 more years, and so he has received little credit for unraveling the causes of the aphasias. In 1873, only 13 years after Broca examined his patient Tan, German neuroscientist Carl Wernicke (1848–1905) evaluated a stroke patient who could speak but could not understand what was said to him. After the patient's death, Wernicke found a lesion in the posterior temporal-parietal region of his brain, and he called it *sensory aphasia*, which is now known as *Wernicke's aphasia*. In 1874, Wernicke (at the age of 26) published a book on types of aphasia and their neurological basis. He called Broca's aphemia *motor aphasia* and postulated correctly that the two regions responsible for motor and sensory aphasia were interrelated (now known as *Broca's area* and *Wernicke's area*; they transmit information through the arcuate fasciculus). Thus, Broca's aphasia involves spared language comprehension but impaired speech, whereas Wernicke's aphasia involves impaired language comprehension but spared speech. Further, it is suspected (but not yet empirically demonstrated) that the former is more likely associated with depression and the latter with paranoia. Wernicke was also an astute observer of clinical symptoms in his patients and described other types of aphasia, such as total aphasia (impaired speech and impaired understanding), and movement disorders, such as ataxia (loss of voluntary muscle coordination). He also described a triad of symptoms (ocular, motor,

Fig. 1.1 Lucy (*Australopithecus afarensis*). Source: Wikimedia Commons.

Fig. 1.2 Mode 1 Oldowan stone tool and flakes. Source: Frederick L. Coolidge.

Fig. 1.3 Mode 2 Acheulean handaxe. Source: Frederick L. Coolidge.

Fig. 3.1 Michelangelo's *Creation of Adam*. Source: Wikimedia Commons.

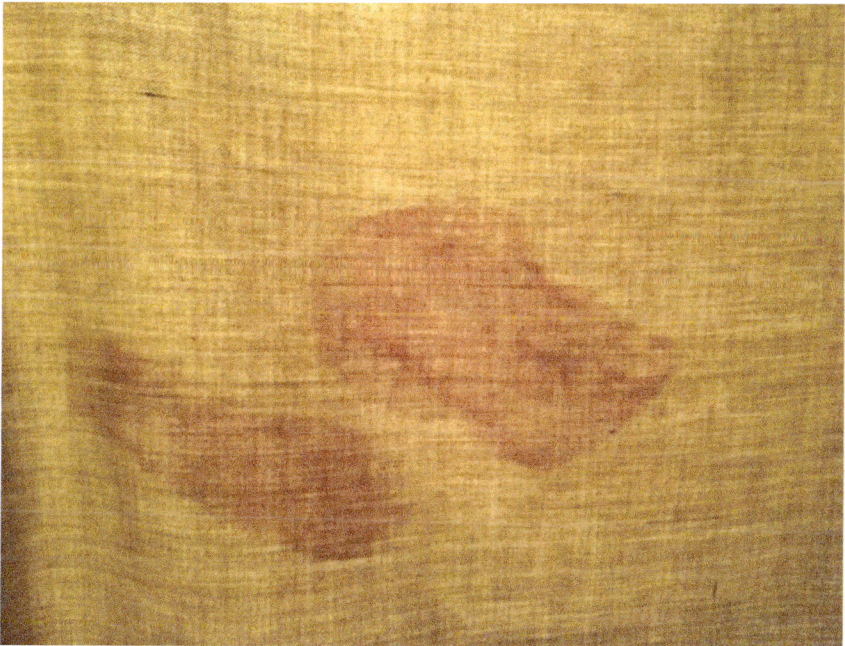

Fig. 6.2 A red wine stain on a shawl. Source: Frederick L. Coolidge.

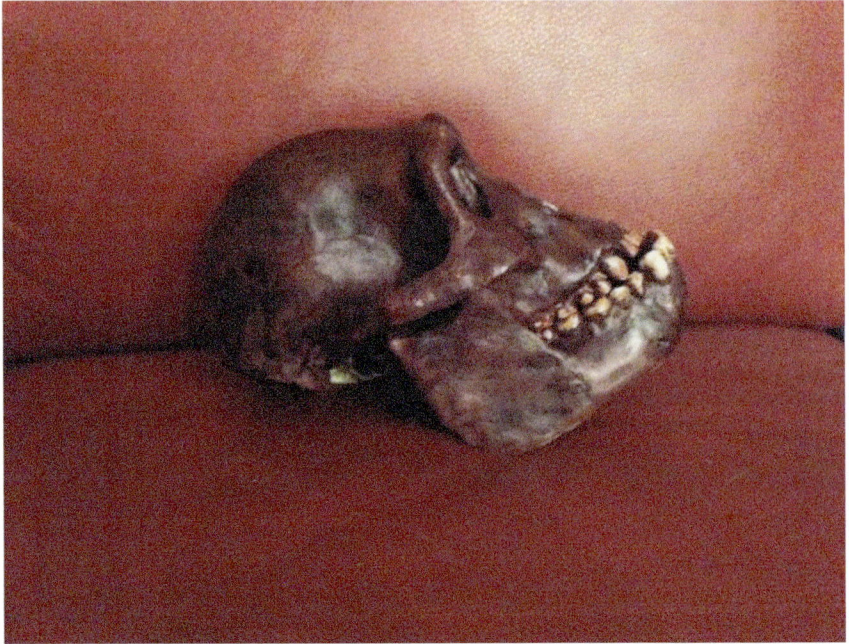

Fig. 6.3 Lucy's skull (*Australopithecus afarensis*). Source: Frederick. L. Coolidge.

and cognitive) in 1881 that are now known as *Wernicke's encephalopathy*. It is thought that this condition is closely related to Korsakoff's psychosis, an amnesic disorder, which the Russian neuropsychiatrist Sergei Korsakoff described in 1887, 6 years after Wernicke published his cases. Both may be caused by a thiamine deficiency in chronic alcoholism, although there are some exceptions (e.g., non-alcoholics and people without thiamine deficiencies).

Temporal Lobes

Superior Temporal Gyrus

The original adaptation of the superior temporal gyrus may have been the application of meaning to sounds, thus the superior temporal gyrus is critical to basic vocal communication and language comprehension. The superior temporal gyrus (BA 41 and 42) includes the primary auditory cortex (also called *Heschl's gyrus*; named after the Austrian anatomist who first identified it), which receives auditory stimuli directly from the ears' cochlea. The posterior portion of the superior temporal gyrus is called the *planum temporale* (posterior BA 22), which is found to be one of the brain's most asymmetrical regions, being larger in the left hemisphere in humans and some great apes such as chimpanzees (e.g., Gannon, Holloway, Broadfield, & Braun, 1998). Because humans and chimps shared a common ancestor from somewhere between 6 and 11 million years ago, this asymmetry suggests that the left planum temporale evolved to be larger even earlier than that. Interestingly, as noted in Chapter 3, symmetry between the left and right planum temporale or asymmetry favoring the right hemisphere has been shown to be associated (but not necessarily causal) with schizophrenia and dyslexia (specifically, right asymmetry in dyslexic boys; Altarelli et al., 2014). It has been proposed that the planum temporale may be viewed a computational sound center that differentiates among various kinds of sounds (spoken, musical, etc.). It may also discriminate among simultaneous sounds, identify their location, and track their movements. In this model, the planum temporale is not only a processing center but also serves as a hub sending identified and discriminated sounds for further processing. Perhaps its asymmetrical nature may also reflect the relative evolutionary importance of auditory (warning cries and other sounds that enhance fitness) and spoken communication in the left hemisphere over tonal recognition (pitch, sound intensity, etc.) in the right hemisphere. The asymmetry may also reflect why modern human brains have been so profoundly influenced by the evolution of primates, who, as small, nocturnal, tree-dwelling animals, had to compete with much larger and more dangerous animals for foodstuff like fruit through vocalizations. Undoubtedly, natural selection acted on any chance mutations that even slightly enlarged regions that processed sounds (the temporal lobes specifically). Even in modern monkeys, the same regions process sounds as in humans; however, it has been amply demonstrated that the temporal lobes in humans have greatly expanded.

The exaptation of the function of the left hemisphere's superior temporal gyrus appears to be the generation of meanings of sounds and preparing them for production by Broca's area. In cases where these meanings are not spoken, this process may simply result in "inner speech." The right hemisphere's superior temporal gyrus is active when the pitch (or frequency) of sounds requires discrimination, such as telling the difference between C and D musical notes and discriminating between sound intensities. As the posterior portion of the superior temporal gyrus is adjacent to the parietal lobes' supramarginal gyrus (BA 40) and angular gyrus (BA 39), forming a junction, it is not surprising that the three regions have intimate language-related functions. This junction, the temporal parietal junction (TPJ; sometimes referred to as the temporal occipital parietal [TOP] junction, or in my state, Colorado, the POT junction) has long been thought to be important to the evolution of language (discussed later in this chapter) and aspects of modern thinking such as social cognition. The TPJ appears to be part of a social brain network along with the amygdala, fusiform gyrus, and orbitofrontal and ventromedial prefrontal cortices. It plays a critical role in imagining one's feelings, needs, desires, and goals and those of others. This imagining process has been called *mentalizing,* and when mentalizing involves others, it is usually referred to as *theory of mind.* The TPJ also performs visual analyses of agency and the perception of animacy (whether an object is alive). Evolutionarily, it was visually important for an organism to recognize prey, predators, or potential mates, and so neurons in the area of the TPJ were selected for their ability to perceive animacy and apply meaning to those perceptions. The latter ability has also been called *social agent detection.* The TPJ appears to be a nexus for social cognition. It is a hub of converging information for language streams of information, attention, and working memory and identification of the social context for all of that information. Again, the TPJ does not act independently, of course; it acts in coordination with some of the previously mentioned brain regions (frontal, parietal, temporal, occipital, and subcortices like the hippocampus, amygdala, and basal ganglia).

Medial Temporal Gyrus

The medial temporal gyrus is an important part of the basic declarative memory system, that is, the recall of words, facts, and details. It is both behaviorally and neurologically distinguished from the procedural memory system, which is the learning and recall (i.e., performance) of physical and motor skills like juggling, stone knapping, playing lead guitar like Eric Clapton, or riding a bicycle. Certainly, all three gyri of the temporal lobes have overlapping functions with regard to language operations. The declarative memory system is generally thought to consist of the anterior (perirhinal and entorhinal cortices) superior gyrus of the temporal lobe, the medial and posterior portions (parahippocampal cortex) of the medial temporal lobe, the inferior gyrus of the temporal lobes, inferior portions of the prefrontal cortex, and the hippocampus.

When confronted with an ambiguous word (or phrase), such as the word *run* when it could be used as a noun or a verb, the medial temporal lobe is initially involved in the basic retrieval of the various meanings of the word. The disambiguation process additionally recruits inferior frontal cortices. It is thought that these inferior frontal cortices help select a particular meaning and then engage the medial temporal lobes again in a second interpretation and unification of meaning, along with recruiting superior portions of the temporal lobes for further integration of the information. If a novel (or foreign) word is heard, the medial and superior temporal lobes and inferior parietal lobes may all be engaged in the active maintenance of that word. If the word is repeated enough times vocally or subvocally and without major distractions, the word will gain obligatory access to long-term declarative memory storage regions. The new word may also gain obligatory access to the long-term declarative memory system if a person is highly motivated or highly aroused. Thus, as noted previously, traumatic events do not need to be repeated to be memorized, as their occurrence often dramatically increases one's state of arousal, which nearly guarantees their near-permanent residence in long-term storage. Some infrequently used words like *anodyne* (which means "inoffensive") are unlikely to create high arousal states, thus they must be rehearsed repeatedly to be memorized.

Inferior Temporal Gyrus

The inferior temporal gyrus is part of the ventral stream of visual processing that originates in the occipital lobes (BA 17) and adjacent occipital cortices (BA 18 and 19) and projects to the temporal lobes. Its primary purpose is object recognition, including shape and color. Medial to the inferior temporal gyrus is the fusiform gyrus (BA 37). The posterior portion of the fusiform gyrus is also adjacent to the anterior portion of the inferior occipital lobe's gyrus. The fusiform gyrus is larger in the middle and tapered at both ends; it gets its name from the Latin word *fusus*, which means "spindle," because of its spindle-like shape. Medial to the fusiform gyrus in the medial temporal lobe is the parahippocampal cortex. As noted in Chapter 3, written words activate the left fusiform gyrus, which is called the *visual word form area* (VWFA; McCandless, Cohen, & Dehaene, 2003) and faces activate the right fusiform gyrus (*fusiform face area,* or FFA). Further, the VWFA is activated in any cultures' writing system, because most writing systems use an average of only three strokes to form letters. It remains uncertain how neurons in the VWFA became so quickly exapted, because writing systems are so recent in terms of *Homo sapiens'* evolutionary history. The oldest of writing systems date back only about 7,000 years, yet fully modern minds (and brains) were highly likely to have evolved by 45,000 years ago. Thus, it seems more probable that the highly specialized and selective neurons of the VWFA were exapted for reasons not directly related to recognizing word forms, but rather were selected for their ability to recognize discrete visual properties of objects and later became exapted to recognize whole words.

Work by neuroscientist Charles Gross (2008) may shed light on this exaptation, as he and his colleagues found that individual neurons and bundles of neurons in the inferior temporal lobes can be modified by experience and directed attention, despite being highly selective to particular properties of visual stimuli (e.g., face cells, hand cells, slits, slots). The latter flexibility may account for the amazing ability of most children in any culture to become highly proficient at reading in a short period of time. In the neuronal recycling hypothesis of Dehaene and Cohen (2007), these cultural acquisitions, like reading, must still find a neuronal niche, that is, a group or network of neuronal structures that can already provide a pre-existing cortical bias for the novel tasks that may be required by a particular culture. Thus, in their view, neural substrates may be engaged in tasks very different from their originally selected function. How this recycling occurs is not clear, but it is without question that neural substrates are put to use in a wide variety of novel cognitive tasks other than those for which they might have been intended. It may be that they are engaged in the new tasks on the basis of their original cortical biases. In other words, it is unlikely that brain neurons and brain regions after birth are pluripotent (i.e., they can support any function), and it is unlikely that they can support or handle any novel task; that is, they are partially constrained by their previous adaptations.

Additionally, but poorly understood, at least at the functional neuroanatomical level, is how the translation between written words and their phonological translations occurs. Dehaene, Cohen, Sigman, and Vinckier (2005) have made the interesting proposal that there is a processing gradient within the inferior temporal cortex that proceeds posteriorly to anteriorly by initially recognizing features of letters (i.e., their local combination detector model), then the letters, then bigrams (adjacent letters), and finally whole words. This sequential processing is thought to continue into the anterior temporal areas until increasingly finer semantic discriminations have been made. The major pathway for this processing appears to be the fusiform gyrus.

The fusiform gyrus can be identified in the brains of the great apes (humans, gorillas, chimpanzees, bonobos, and orangutans). However, there is a cytoarchitectural (cellular shape, complexity, and size) asymmetry present in humans' fusiform gyri that is not present in chimpanzees (our closest genetic relative) or in the other great apes. This cytoarchitectural asymmetry and even cerebral hemisphere asymmetry for language areas in the left hemisphere of humans, which is absent or much less prominent in chimpanzees, appear to support the rather obvious hypothesis that humans have a much richer and exacted ability to understand and make discriminations in their visual environments than any other extant hominoid (e.g., Chance, Sawyer, Clover, Wicinski, Hof, & Crow, 2013; Weiner & Zilles, 2016).

Neuropsychopathology

Prosopagnosia

Some fascinating neuropsychological symptoms are associated with inferior temporal lobe dysfunction. The inability to recognize faces, called *prosopagnosia*, had been attributed to lesions or disturbances in the fusiform gyrus, usually in the right hemisphere's fusiform gyrus. Human brain studies by

Fig. 6.2 Red wine stain on a shawl (See color plate).
Source: Frederick L. Coolidge.

Rangarajan et al. (2014) support previous research that the left and right hemisphere's fusiform gyri are both involved in face perception but with subtly different roles. Electrical brain stimulation of the right fusiform gyrus caused a distortion of faces whereas stimulation of the left fusiform gyrus produced nonspecific visual changes, like distortions of color and points of light.

The fusiform gyri's neuronal evolutionary penchant for seeing faces may help to explain the phenomenon of *pareidolia*, in which faces are seen in inanimate things like clouds, rocks, windows, tacos, and even wine stains on shawls. The latter happened to me one night while watching TV. I spilled some red wine on a shawl, and in the morning when I held up the shawl, I saw what is represented in Fig. 6.2.

No doubt because of my interests, I saw Lucy's 3.1 million–year-old skull in the stain on the right. I went to my display case, pulled out my replica of Lucy's skull, and took the picture presented in Fig. 6.3.

When one of my graduate students at the Indian Institute of Technology Gandhinagar used these two pictures for a poster presentation on pareidolia, we were amused yet again to see the face of a bushy bearded man on the

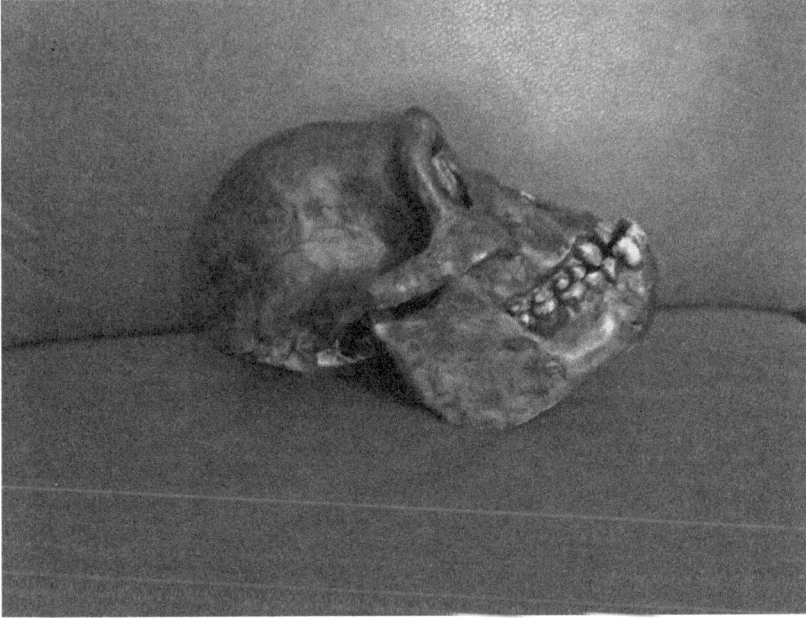

Fig. 6.3 Lucy (*Australopithecus afarensis*) (See color plate).
Source: Frederick L. Coolidge.

right temporal region of Lucy's skull—Charles Darwin noted the importance of being able to recognize faces and facial expressions for successful social interactions. Also, successful recognition of facial expressions in fellow human beings must have been crucial (enhanced fitness) in a wide variety of social situations such as mating, parenting, and trade. Perhaps the evolutionary cost of failing to read successfully the face of another person was greater than the perception of a person's face in an inanimate object.

Synesthesia

The word *synesthesia* comes from the Greek and means "joined perception." *Synesthetes* are people stimulated in one sense who are simultaneously stimulated in another sense. One of the most common forms of synesthesia is grapheme-color synesthesia (graphemes are individual numbers or letters). Grapheme-color synesthetes reliably see the same color superimposed on particular numbers or letters. Interestingly, letters and numbers are the primary inducers of synesthesia, followed by music and sounds. Color is by far the most common perception in synesthesia. Synesthesia appears to have a genetic origin but does not appear to be sex or gender specific. While prevalence estimates of synesthesia vary widely, one current estimate is about 4 to 5% of all people. One neurocognitive model that attempts to explain grapheme-color synesthesia proposes that regions of the brain that perceive color in the primary visual cortex (BA 17), and adjacent exstriate occipital cortices (BA 18 and 19) "leaks" that information into the fusiform gyrus in its perception of numbers or letters. Other models implicate the superior temporal gyrus (disinhibition) and the angular gyrus (BA 39; there also may be cytoarchitectural differences in this region between synesthetes and non-synesthetes).

Ideasthesia

An interesting related phenomenon is *ideasthesia*, in which perceptions are paired with concepts but in a non-arbitrary manner. In 1929, cognitive psychologist Wolfgang Köhler presented to Tenerife islanders (primarily Spanish speaking) two figures and two sounds and asked them to associate each sound to only one of the figures. Köhler found a clear preference, although it appeared that the sounds and figures were completely arbitrary. In 2001, Ramachandran and Hubbard repeated the experiment, using the shapes shown in Fig. 6.4.

Which of these figures would you label *bouba* and which would you label *kiki*? Ramachandran and Hubbard found that American college students and those who speak Tamil (a language spoken in southern India and Sri Lanka)

Fig. 6.4 The bouba/kiki effect.
Source: Wikimedia Commons.

overwhelmingly associated *kiki* with the figure on the left and *bouba* with the figure on the right. This preference has been shown in children as young as 2½ years old, although at least one study has shown the preference may be absent in autistic people. Ramachandran and Hubbard suggested this finding shows that the objects' names might not be completely arbitrary, and in some way, it may be a form of synesthesia. It is also possible that the "kiki/bouba effect" is simply a form of priming (experience with a stimulus influences later responses), as pronouncing *bouba* requires one's mouth to be rounded, while pronouncing *kiki* requires a more horizontal, angular mouth shape.

Ramachandran and Hubbard further proposed that this linking of sounds and perceptions might form a basis for the evolution of language. The evolution of language will be dealt with in greater depth later in this chapter, but their hypothesis is that language began with a kind of protolanguage, that is, a vocal communication system among early hominins in which objects were named by contours of the mouth in a form of synesthesia. Although they claim that this process alone would be insufficient to account for fully modern syntactical language, they propose that, if boot-strapped with other related processes like sound contour and vocalizations, disgusting smells, and facial expressions, the resulting coalescence of seemingly minor processes might still be a basis for a protolanguage.

Action Perception Circuits, Neural Reuse, and Language

Recently, language neuroscientist Friedmann Pulvermüller (2018) highlighted the neuroanatomical principle of the rich connectivities of neurons as well as their more specific long-distance connectivities. For the emergence and origins of language, he found that cortical areas linking sensory and motor neurons were particularly relevant for the foundations of language. These areas, as noted in Chapter 5, were labeled *convergence zones* by Damasio (1989), and they are also referred to as *connector hubs* (Gordon et al., 2018). Pulvermüller (2018) noted that motor neurons receive and process information from sensory neurons and vice versa. In this manner, motor and sensory neurons influence each other's information processing, which establishes what Pulvermüller called "action perception circuits" (APCs). The formation of APCs is influenced by these sensorimotor operations, and it is their co-activation that strengthens their interactive influence. The mechanisms associated with APCs thus provide a neuroanatomical substrate for the gradual emergence of language. According to Pulvermüller (2018), there is a "neural reuse of action and perception mechanisms for language concepts, and communication. . . . Action perception circuits provide a basis for higher cognitive mechanisms, including attention, memory, predication and combination, and how they interact in linguistic and semantic/prag-matic processing" (p. 36). Pulvermüller hypothesizes that APCs not only map the environment on neural substrates, but their intrinsically linked iter-ative computations of action and perception allow for the emergence of ab-straction and creativity. The co-activation of APCs may also provide "a brain basis for combination thus capturing syntactic, semantic, and pragmatic in-formation. Symbol-related APCs also link object and action representations in the brain, which yields a mechanism for semantic grounding" (p. 37). This combinatorial semantic learning also provides a basis for intentional communicative actions, which Pulvermüller views as a basis for pragmatic speech acts, that is, the reasons for and purposes of speech.

The Evolution of Language and Protolanguage

Language evolved via natural selection for the primary purpose of commu-nication, and it began in a more simplistic form (i.e., a protolanguage). It is absurd to assume it (1) arose whole cloth from a simple genetic event in one person about 100,000 years ago, and (2) its primary purpose was *not* for com-munication, as Chomsky and his colleagues have contended (e.g., Bolhuis, Tattersall, Chomsky, & Berwick, 2014). From only these two sentences, the reader can already tell that the nature and origins of language are highly con-tentious issues, and perhaps it is a bit of a truism: All linguists hate other linguists. However, this is perhaps overstated: Linguists often strongly differ in their opinions about language's origins and its initial nature.

Let us start this discussion with American linguistics professor Ray Jackendoff's (1999) hypothesis that all modern language functions devel-oped incrementally, originally through the cortical auditory-vocal networks of primates. As has been noted throughout this book, human brains have largely been informed and influenced by primate brains (e.g., Aboitiz, Aboitiz, & Garcia, 2010); furthermore, human adaptations and exaptations are constrained by common ancestors. Again, these small nocturnal pri-mates, when they diverged from the mammalian lineage about 65 million years ago, managed to compete with other animals for fruit by coordinating their activities through other-directed vocal cries and a common perception of those cries. Varying cries are easily learned through associative learning (i.e., classical and operant conditioning). The original purpose of these cries might even have been *intrapersonal*, that is, only for the vocalizer. As noted earlier, the reasons why organisms use sounds or why humans use speech are called the "pragmatics of speech." It has been proposed that there are five major pragmatics of speech: exclamatives, imperatives, declaratives, interrogatives, and subjunctives. I hypothesize that the first pragmatic of speech was exclamatives, which expressed emotions, and in these first primates, it might have been to express pain or pleasure or to exert dominance. As stated earlier, I think these exclamatives (particularly painful ones, like "ouch") were *not* to warn others but simply to express one's emotional state (in this case, pain). Quickly, however, hearing another primate cry out in pain would become an *interpersonal* communication, as one might understand that the cry was associated with pain (an example of adapted empathy). This use of exclamatives then might serve as a valuable warning of thorns, poisons, or predators. Exclamatives, therefore, are generally short bursts of speech that do not require complicated brain networks for their expression or comprehension; they are based on an auditory-vocalization brain network. A second pragmatic of speech might have been imperatives, that is, commands, as social hierarchies are characteristic of all extant human and nonhuman primates. So vocalizations to move, submit, share food, or drop food were likely to have been emitted by those higher in the social hierarchy to those lower in it (because of sex, size, age, or innate aggressiveness).

Before elaborating on the third pragmatic of speech, declaratives (naming things and actions), it is important to point out a common fatal flaw in some linguists' and anthropologists' arguments. Some have suggested that a protolanguage must have had the voluntary use of *symbolic* vocalizations; that is, the first users of a protolanguage must all agree that a particular arbitrary sound will now stand for some arbitrary thing or action (e.g., Deacon, 1996). This flawed position further implies that the group's members have met and agreed on a specific meaning of an arbitrary sound. While this sounds rather rational, it is not. The problem is Occam's razor: A simpler explanation is preferred over a more complex one, or the explanation with the fewest assumptions is most likely. Classical and operant conditioning, which can be demonstrated in simple, almost brainless organisms like planaria, can explain the association of some arbitrary sound with some thing or action. I will again use Pavlov's dog as an example, where a previously arbitrary (neutral) stimulus (the sound of a bell) is paired repeatedly with food (said to be an unconditioned stimulus because a dog does not need to "learn" to like food; the attraction is innate). The food is said to be an unconditioned stimulus because of its smell, and its visual presence *elicits* an automatic response from the dog (salivation and arousal). Again, this automatic response does not have to be learned. All kinds of organisms, complicated or uncomplicated, have built-in or automatic responses (reflexes) to a limited number of stimuli. If paired enough times, the neutral stimulus will elicit a response from the dog (the neutral stimulus becomes a conditioned stimulus when it finally elicits a response). Thus, discussions of symbolism or symbolic vocalizations must be cautious, otherwise, some might be tempted to say the bell has become a symbol of food for the dog, where the process of association between the bell and food is both cognitively and neurologically simpler. Therefore, without invoking Deacon's complex, shared symbolic vocalization hypothesis, things or actions may be reliably associated with particular sounds through simple conditioning, which serves as the cognitive basis for declaratives. Of course, declaratives as a pragmatic of speech can become much more complex as brains and cognition become much more complex, so these declarative statements of fact can be used to manipulate others and lie to others, and be put to other nefarious uses. It is provocative to ponder whether the subsequent manipulative use of declaratives was an exaptation while the simple associative learning of sounds with specific things was the adaptation. However, in the origins of a protolanguage, declaratives may have simply resembled the "vocabularies" of modern vervet and putty-nosed monkeys. Also, once declaratives have been established, interrogatives (questions) could have followed (which is my fourth pragmatic of speech). So not only could a specific sound be associated with a specific thing like a fruit, but an early hominin using a protolanguage could ask where a fruit came from.

At this juncture, I have established four possible pragmatics of a protolanguage: exclamatives, imperatives, declaratives, and interrogatives. Importantly, I have done so without having to rely on the constructs of symbols, symbolizations, or symbolic thinking. Further, it is interesting to note that all four of these pragmatics of speech do not require more than a single syllable (at least in English), like "Ouch! Move! Deer! Where?" This syllabic simplicity might allow cognitively simple organisms to have a protolanguage, which consists of a limited range of pragmatics and a large vocabulary. The latter characteristic is possible owing to the virtually unlimited learning possibilities afforded by associative learning (e.g., Bronfman, Ginsburg, & Jablonka, 2016).

As noted earlier, Chomsky and his colleagues (Bolhuis et al., 2014) have claimed that language appeared suddenly and whole cloth, in one person. They also claim it was not originally for the purpose of communication and was not subject to natural selection. However, the evidence used by a clear majority of linguists, anthropologists, language geneticists, and evolutionary psychologists almost inescapably reveals an opposite position: language evolved *slowly* by way of protolanguages, evolved for *purposes of communication*, and *was subject to natural selection* (i.e., those who could communicate more efficiently and effectively were more likely to be successful at reproduction; greater fitness). However, it is possible that something "suddenly" happened to language about 200,000 to 100,000 years ago that led to fully modern thinking and fully modern language. And this leads me to my fifth and most cognitively complex pragmatic of speech: the subjunctives. Subjunctives express hypothetical states, events or situations that may not have occurred or express ideas contrary to fact. Sometimes, they may express what people can imagine or what they wish to be true. However, subjunctive thinking may have required particular cognitive prerequisites, as will be discussed next.

Cognitive Prerequisites for the Subjunctive Pragmatic of Speech

In two articles, Chomsky and his colleagues (Fitch, Hauser, & Chomsky, 2005; Hauser, Chomksy, & Fitch, 2002) proposed that the hallmark of modern human language is recursion. *Recursion* is the embedding of a phrase in another phrase such that the embedded phrase modifies or adds to the original meaning of the phrase (for a detailed discussion of recursion, see Coolidge, Overmann, & Wynn, 2010). As noted earlier, Bolhuis et al. (2014) have essentially argued that language did *not* evolve via natural selection, and fully modern language, which includes recursive phrasing, appeared suddenly about 100,000 years ago. Also, as noted previously, they argue strongly against any kind of protolanguage, although their thinking in this regard is held by only a small minority of linguists (for detailed discussions of protolanguage and issues with Chomsky and his colleagues' contentions, see Bickerton, 2007; Botha, 2002; Gärdenfors, 2013; Hurford, 2014; Jackendoff, 1999; Pinker & Jackendoff, 2005). Evolutionary linguist James Hurford (2014) has attempted to quantify these two different positions from five perspectives. He characterizes the Chomskyian position as claiming that a single genetic mutation caused the sudden appearance of fully modern language with recursion, language itself constituted a completely new cognitive domain, language was primarily for internal use and not communication with others, and language was not subject to natural selection. Hurford and a majority of other contemporary linguists take the opposite position, although he notes that neither overall position is absolutely correct (except that the single genetic mutation, all-of-a-sudden appearance of language in one person is ridiculously incorrect). The majority position (which I endorse) is that a succession of mutations over millions of years, whose phenotypes were subject to natural selection, resulted in increasingly more complex communications with other organisms. It did not result in a distinctly new cognitive domain requiring specialized neural circuitry, but rather arose from pre-existing cognitive domains and neural circuits, like those involved in basic associative learning. Another pre-existing cognitive domain may have been the ability to store sounds (phonological storage). Over time, language's utterances also became increasingly more complex, and one result of the complexity may have been greater reproductive fitness.

Did Something About 100,000 Years Ago Change the Course of Language Evolution?

There may be a modifying position between these two schools of thought about language's sudden appearance 100,000 years ago. I stand firmly with the linguists who believe language became increasingly complex over time (the essence of a protolanguage theory). However, archaeological evidence does suggest that there was a cultural explosion beginning about 50,000 years ago, attributed to the *Homo sapiens* who left Africa about 80,000 to 50,000 years ago. As noted earlier, their culture is called the Aurignacian, and it included elaborate personal ornaments, depictive cave paintings, creative figurines, and highly ritualized burials. There are some tantalizing hints that *Homo sapiens*, perhaps beginning about 200,000 to about 160,000 years ago, may have already carried the cognitive abilities to handle basic language (as *Homo idaltu* (another early hominin), about this time, appeared to have carried around polished pieces of human skulls; White et al., 2003). However, the explosion of culture more recently than 100,000 years ago suggests to me and my colleague, archaeologist Thomas Wynn (Coolidge & Wynn, 2005,

2009; Wynn & Coolidge, 2010), that something did happen to language around this time, and we think there may have been some beneficent mutation that affected one of the cognitive domains necessary for language. Yes, most DNA mutations are harmless or uneventful, while others are harmful or deadly. Despite their relative rarity, however, beneficent genetic mutations which enhance fitness do occur. One possibility for our enhanced working memory hypothesis is that a mutation (or epigenetic event) increased phonological storage capacity. As noted earlier, the four pragmatics of speech (exclamatives, imperatives, declaratives, and interrogatives) can all be handled in most extant languages with short sounds (single syllables in most cases). Jackendoff (1999) has referred to them as "the one-word fossil stage of language evolution" (p. 273). And if putty-nosed monkeys can combine individual utterances to form a distinctly different meaning, then it is reasonable to assume *Homo sapiens* could have had an elaborate vocabulary about 100,000 years ago (and probably *Homo erectus* had an extensive vocabulary 1.9 million years ago). However, phonological storage capacity limits how much one can keep in one's mind in order to be spoken. Here again, the Chomskyites (i.e., followers of Chomsky) might be partially correct: internal dialogue or internal thinking probably preceded external public language. So a beneficent mutation that increased phonological storage capacity may have been the necessary and critical cognitive prerequisite for the subjunctive pragmatic of speech because it almost always occurs in language in a subordinate clause (a definitive characteristic of recursion) and is usually introduced by a conjunction (and, but, then, etc.).

Evolution of Grammar and Syntax

Another possibility is that a beneficent mutation affected general working memory capacity, which would have the downstream effect of increasing phonological storage capacity, as well as enhancing the capabilities of the visuospatial sketchpad and the episodic buffer. However, often neglected in nearly all discussions of protolanguages is the evolution of grammar and syntax. *Syntax* usually refers to the rules for ordering words in sentences. *Grammar* also has this connotation, but it is also a more comprehensive term, as it encompasses word order, the use of inflections, phonology (relationships between sounds), and semantics (cognitive structures of meanings). Others have noted that rules for word order may have evolved to minimize online memory requirements (e.g., Ferrer-I-Cancho, 2015). I have similarly argued (Coolidge, 2012) that syntax may have evolved specifically to bypass the limitations of working memory capacity and/or phonological storage limits. If all sounds could arbitrarily be associated with things (nouns, whether actors or acted upon) or actions (verbs), then a protolanguage would have inherent limitations, as the central executive component of working memory would have to parse out whether a sound was an actor (subject, S), an action (verb, V), or something being acted on (object, O). With a culturally imposed word order, working memory would be spared this initial parsing requirement, as an individual would have learned that in their culture the actor comes first in a sentence, the action comes second, and the object or whatever is being acted on comes last. At first glance, word order appears relatively arbitrary, as it would not matter whether the action was first, second, or last, if all of its members agreed on a specific order. However, the limits of working memory capacity and/or phonological storage are perhaps the major influence on a culture's choice of word order. As Ferrer-I-Cancho (2015) has noted, all languages and their speakers are constrained by "online memory" limits. Interestingly, nearly all extant languages have an imposed word order, and more speakers use the SVO word order (like English), whereas and more languages use the SOV order. OVS and OSV are rare, which probably indicates that greater demands are placed on working memory capacity in those orderings, although other factors may be involved. For example, the fact that more languages revert to SVO rather than SOV could be the result of cultural imperialism, that is, the imposition of a word order preference by a culture on another culture, according to Ferrer-I-Cancho. Nevertheless, I would hypothesize that the tendency for verbs to revert to the center of a phrase (rather than regress to the end) indicates that that placement makes the fewest demands on working memory capacity, which I view as the primary influence on word order. Further, with capacity-limited working memory freed from parsing each sound's placement in the sentence and with verbs' natural reversion to the center, it might free working memory for higher-order syntactic calculations like recursion. Still to be determined is why the SOV word order, dominant among languages, is viewed as seminal and the most likely ordering among protolanguage and ancient languages. Further, what factors influence the regression of SOV to SVO, why is the opposite is so rare (SVO to SOV), and what selective advantage comes from the movement of V to the center of a word order? I agree with Evans and Levinson (2009) that language design has multiple influences and constraints; however, as

I noted earlier, I suspect that limited working memory and/or phonological storage capacity are primary and dominant reasons for word order and, thus, a strong impetus for syntax and grammar.

Other Purpose of Language: Gossip

While the word *gossip* appears to have negative connotations, the literal meaning involves a sharing of information among close friends or close relatives. Anthropologist Robin Dunbar (2004) has hypothesized that, given the strong evolutionary constraint on *Homo sapiens*' evolution by the highly developed early primate social behavior, verbally exchanging information (gossip) became a means of bonding in large social groups. When the genus *Homo* (specifically, *Homo erectus*) made the full transition to terrestrial life by about 1.9 million years ago, larger social groups were necessary to avoid predation and to minimize the risk of exploring new territories for food and other resources. However, the increase in social group size was not without a social cost: Individuals were subject to being victimized by more dominant members, individual gains might be sacrificed for the greater good of the group, and some individuals could take advantage of the anonymity afforded by a larger group and not work as hard or as effectively as other members. Dunbar labeled the latter individuals *free riders*, and he proposed that gossip provided a means of identifying such members and keeping their behavior "in check." If they were not kept in check, then at the least their behaviors would be revealed to others, and likely the chances of others being exploited by them would be reduced (at least the members involved in that gossip). Thus, rather than seeing gossip as some idle and useless social endeavor, Dunbar views it as reflective of our social primate heritage and a nearly necessary behavior for successful life in a large social group.

Another Purpose of Language: Surprise and Humor

A man at the dinner table who was being handed fish dipped his two hands twice in the mayonnaise and then ran them through his hair. When his neighbour looked at him in astonishment, he seemed to notice his mistake and apologized: "I'm so sorry. I thought it was spinach."

—Sigmund Freud (1905/1960, pp. 138–139)

This joke was reported to be among Sigmund Freud's favorites, and when his friend Wilhelm Fliess suggested a similarity between jokes and dreams, Freud was stirred to write one of his classic works, *Jokes and Their Relation to the Unconscious* (1905/1960). Just like dreams, accidents, and slips of the tongue, which Freud thought revealed unconscious thinking, Freud thought jokes had the potential to expose ideas from the psyche that would otherwise be unavailable to conscious processing. Freud also proposed that jokes could serve as a defense mechanism to divert emotional responses such as aggression and sexual tensions, among other issues.

It is important to note that Freud's spinach joke contains a strong element of surprise, as the ending of this rationally proceeding joke ends irrationally. Anthropologist Elliot Oring (1992) uses the phrase *appropriate incongruity*, defined as the condition in which some parts of a joke are normal or appropriate and other parts of the joke are weird, strange, or incongruous (inappropriate or not consistent). When these parts of the joke are joined together, it creates an *appropriate incongruity*, which interestingly, humans often *naturally* find funny. Some have suggested that it is the surprise element that helps make the joke funny, at least initially (e.g., Dessalles, 2010). This surprise element may derive from early childhood "surprise" games, like peek-a-boo, that invariably and cross-culturally make children laugh. Thus, when a person says, "A kangaroo walks into a bar . . . ," listeners are initially surprised and expectations may be set for an even more surprising ending.

Psychologists who study humor have long noted the element of surprise in jokes, but they do not often speculate as to why surprise is appreciated. Linguist Jean-Louis Dessalles (2010) has noted that people tend to like other people who surprise them with either a recent news event, with juicy gossip, or with a joke. He hypothesizes that humans do so because evolutionarily it was always important *not* to be surprised by other people, animals, or events. Thus, Dessalles propose that people tend to surround themselves with other people who may be able to alert them to hidden dangers, and gossip about others and jokes serve as two of these alerting mechanisms. According to Dessalles and others, people may now appreciate gossip and jokes that have an element of surprise (and the people who tell them) because people unconsciously favor anyone who is able to anticipate danger.

Recently, it has been argued (e.g., Kozbelt, 2019) that humor and creativity may be inextricably linked, as they are both valued by others and are most often viewed as desirable characteristics in future mates. Interestingly, empirical evidence from dating studies find that men and women have significantly different values about humor in future mates, reflecting differing mating strategies. Men have been found to value humor receptivity in women as essential, while they view humor production in a woman as a kind of bonus. Women tend to see humor production as essential in a mate, especially in assessing a long-term relationship, although humor receptivity was also valued. Empirical studies have also found humor production to be correlated with general intelligence and other cognitive abilities, including verbal creativity. Kozbelt suggests that humor may serve as a relatively safe indicator of compatibility between two people, as the success of a joke often depends on implicit values and attitudes of producer and the receiver. For example, a hostile, aggressive, or demeaning joke will fall flat unless the producer and receiver of the joke have similar attitudes, and it has been shown that couples with similar senses of humor tend to stay in their relationships longer than couples with dissimilar senses of humor.

Summary

1. One of the major adaptations of the temporal lobes is the storage and application of meanings to sounds.
2. With the evolution of primates and later hominins, the temporal lobes became exapted for protolanguage and later fully modern recursive language.
3. Language evolved by means of natural selection for communication.
4. A recent beneficent genetic event may have increased working memory and/or phonological storage capacity, which allowed the release of recursive phrasing, a prerequisite for the most complex pragmatic of speech, the subjunctive mode.
5. Language use may have been influenced by gossip, which may have served the function of keeping cheaters and loafers in check in larger social groups.
6. Freud proposed that jokes may have arisen to relieve aggressive and sexual tensions; others have proposed that surprising and incongruent jokes may have emerged to reduce the probability of being surprised. Humor may also be correlated with creativity, and sexual differences in humor production and receptivity have been demonstrated, perhaps reflecting evolutionary mating strategies.

7

The Cerebellum

As noted in Chapter 3, the cerebellum has long been associated with the smooth timing and coordinated execution of motor movements (Fig. 7.1). The ancient Greeks, by the fourth century BC, had anatomically differentiated the cerebrum from the cerebellum (meaning "little brain" or "lesser brain" in Latin), and remarkably there were suspicions that the cerebellum was associated with faster-running animals because of the "intricately folded" layers in the cerebellum (Finger, 1994). Aristotle may have been the first to identify and name the cerebellum, around 350 BC. In an irony associated with its name (little/lesser brain), the cerebellum is thought to contain about 80% of the brain's 86 billion to 100 billion neurons (e.g., Herculano-Houzel, 2016). The cerebellum consists of two major types of cells, Purkinje and granular, and the dendritic branching in the cerebellum's Prukinje's neurons is the most extensive and complex in the entire brain. Interestingly, if just the surface area of the cerebellum were flattened, it would be greater of the surface areas of either cerebral hemisphere. It would also have a surface area about 15% greater than a 20th-century record album cover. Herculano-Houzel found that the cerebellum's neurons are smaller than those in the cerebral cortex, because the cerebellum's neurons have far shorter distances to travel. She also found that the absolute number of total human brain neurons is higher than that in most any animal except elephants, which she determined have about 250 billion neurons. Interestingly, about 98% of their neurons are in cerebellum, presumably to coordinate their functionally complex trunks.

It was not until the Renaissance that brain anatomists began to pay serious attention to the cerebellum, when, in 1573, Italian anatomist Constanzo Varolio correctly identified the structure that connected the cerebellum to the upper spinal cord. Because that structure appeared to him as if it were a bridge, he labeled it the *pons* (Finger, 1994). All afferent (incoming) and efferent (outgoing) neuron bundles travel through the pons–cerebellum connection. As noted in Chapter 3, in 1664, English anatomist and psychiatrist Thomas Willis proposed that the cerebrum was the organ of thought, with the cerebellum and pons responsible for involuntary movements. Other

Evolutionary Neuropsychology. Frederick L. Coolidge, Oxford University Press (2020).
© Oxford University Press.
DOI: 10.1093/oso/9780190940942.001.0001

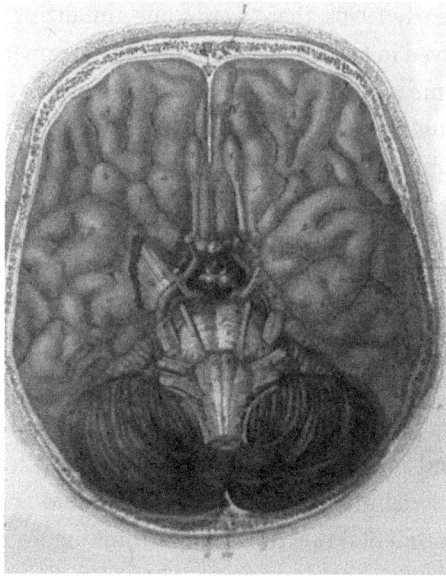

Fig. 7.1 Cerebellum (1890 medical print).
Source: Frederick L. Coolidge.

studies in the 1700s and 1800s further confirmed the role of the cerebellum in motor functions, such as their smooth execution and proper timing, and its importance in maintaining proprioceptive equilibrium (the body's balance). British neurologist Gordon Holmes (1917, 1922) published studies of patients with tumors of the cerebellum and of World War I soldiers with cerebellum wounds. He clearly established the cerebellum's importance to balance, gait, and the smooth execution of gross and fine motor movements, although, as observed later, there was no mention of cognitive or mental problems in these patients. Further, Holmes (1939) may have set up a bias that was to persist for decades, that the cerebellum was not involved in any conscious experiences. He wrote, "I have no doubt, however, that in man even extensive lesions of the cerebellum involve no form of conscious sensation" (p. 7).

In his classic book, *Higher Cortical Functions in Man*, Russian neuropsychologist Alexander Luria (1962/1966) had but a single mention of the cerebellum, with little or no mention of its function, and even a popular article about the cerebellum, in *Scientific American* (Llinás, 1975), mentioned it only as a "central control point for the organization of movement" (p. 66).

By the 1980s and early 1990s, there were some tantalizing hints that the cerebellum was involved in tasks other than muscle movements, such as timing judgments and some aspects of verbal memory. In 1986, Leiner, Leiner, and Dow, based on clinical studies in humans and animal studies, noted that the "human cerebellum may contribute to a broader spectrum of functions than is usually attributed to it ... the connections with the prefrontal cortex could allow the cerebellum to extend such programming [hand or speech muscles] to the ideational manipulations that precede planned behavior" (p. 450). In later papers (1989, 1991), these same authors more strongly stated their hypothesis that the cerebellum was involved in functions other than motor movements, such as cognitive and language skills. In 1993, Japanese neuroscientist Masao Ito proposed that the cerebellum might be likened to a dynamic learning and memory system that automates a learned movement, and he also speculated that this control system could apply to any type of neural operation or mental function such as thinking. In 2002, Ito was even more specific (and prescient):

> Furthermore, I noticed that the control system operation [of the cerebellum] can be identical for voluntary movement and thinking, either verbal or nonverbal, in the sense that while a voluntary movement manipulates a motor plan informed by muscles and bones, thinking manipulates images, concepts, or ideas encoded in the temporoparietal cortex. The control system schemes .... for movement should apply to thought processes for which the prefrontal cortex acts as a controller of images, concepts, or ideas generated in the temporoparietal cortex as control objects. While thought processes are repeated, a model of images, concepts, or ideas created in the association cortex will be copied by the cerebellum so that the thought process can be performed in a feedforward manner. One would be able to think without requiring conscious attention to the feedback of the outcome or even to the action of thinking itself. (p. 283)

A host of neurophysiological studies over the past decades have substantiated Ito's bold hypothesis. The cerebellum has now been implicated in a wide variety of cognitive functions, including insight, intuition, creativity, novel problem-solving, language, semantics, grammar, metalinguistic skills, working memory, verbal fluency, reading, writing, and the rate, force, and rhythm of actions.

Phylogenetics

Phylogenetics is the study of a trait's evolutionary history, using comparisons among species. The cerebellum appears to have evolved in the vertebrate line of the chordates (who possessed a cartilaginous skeletal rod running the length of their bodies), thus, the cerebellum can be traced to the earliest fishes, amphibians, and reptiles, perhaps 400 million years ago or earlier. I have stressed throughout this book that brains evolved primarily for moving, and as I have noted, the cells that specialized in analyzing sensory cell information and then directing motor cells for appropriate and efficient movement became the brain's neurons. However, it appears that evolution favored the development of a "second brain," the cerebellum, one that could automatically take over the analytics of multiple sources of sensory information, and integrate and process that information relatively independently of the bigger brain, yet act in a highly complementary manner. The cerebellum, after completing this sensory information analytic processing, passes this refined or recomputed information back to upper brain regions for reprocessing (e.g., parietal and temporal lobes) and for their final execution (i.e., frontal lobes and prefrontal cortex). The overall process also appears to be highly iteritive. Further, as Ito and others have emphasized, the cerebellum performs its functions whether the information is motor, nonverbal, or verbal, like ideas, concepts, and images. Thus, movements and thoughts are treated identically by the cerebellum.

It has also been suggested that the evolutionary expansion of the cerebellum in *Homo sapiens* and its exaptation for higher cognitive functions did not occur in complete isolation but in tandem with the concerted evolution of the prefrontal cortices, other areas of the cerebral cortex, and motor systems. Therefore, natural selection may not have been just acting on individual brain regions (mosaic evolution) but on functional and interconnected brain systems (concerted evolution). As noted previously, the first stone tool-making technologies, dating to at least 3.4 million years ago, required a concerted effort of various but specific brain regions, including sensory recognition areas, the prefrontal cortex, motor systems, and the cerebellum. Their concerted evolution may also help to explain how subsequent higher-level cognitive functions like language may have arisen from the exaptation of multiregional brain networks already in place for other functions.

Concerted Evolution of the Cerebellum and the Neocortex?

The distant ancestors of modern humans, the hominidae or great apes, date back to 20 million years ago. As mentioned in earlier chapters, the general evolutionary trajectory of anthropoid primates had been for a concerted expansion of both the cerebellum and neocortex. At the time of the common ancestor for humans and great apes, our lineage began to deviate significantly from the general primate trend and move toward a dramatic increase in cerebellum size relative to the neocortex (Barton & Venditti, 2014). There is also some provocative evidence that this trend may have accelerated in *Homo sapiens*, as it has been shown from endocasts that the cerebellum-to-cerebrum ratio is significantly larger in *Homo sapiens* than in Neandertals (Bastir et al., 2011; Kochiyama et al., 2018; Weaver, 2005, 2010).

Since brain matter does not fossilize, how can cerebrum/cerebellum ratios be determined? Paleoneurologists use three-dimensional (3D) methods and scanning of skulls, coupled with sophisticated multivariate statistical techniques, to estimate not only volumes of brains within skulls and endocasts but also their shape. On the bottom of the inside of a skull, there are three major fossa (depressions): anterior, middle, and posterior, which form bases for the frontal lobes, temporal lobes, and cerebellum/pons, respectively. The posterior fossa possesses amazing power for predicting the size of the cerebellum housed within it (i.e., $r = 0.88$, which indicates a very strong predictive relationship; Kubo et al., 2014). Thus, based on the difference in size and depth (Cabestrero-Rincón, Balzeau, & Lorenzo, 2018) of the posterior fossa of skulls of *Homo sapiens* and Neandertals, it is likely that the trajectory of a larger cerebellum relative to the neocortex continued in *Homo sapiens* but not in Neandertals. Further, as noted earlier, the dendritic branching of the Purkinje neurons in the cerebellum is the most complex in the whole brain, and if *Homo sapiens* did have more complex social interactions in their greater-sized groups than Neandertals, then the greater information processing obtained from a larger cerebellum would not be surprising.

In human and nonhuman primates, the anterior portion of the cerebellum controls motor movements, but it appears that the lateral and posterior portions of the cerebellum expanded in size and scope of function in *Homo sapiens*. The lateral and superior posterior portions of the cerebellum are so phylogenetically younger and distinct that it has been labeled the *neocerebellum* (Schlerf, Verstynen, Ivry, & Spencer, 2010). These newer regions of the cerebellum are still involved with the execution of motor movements, but they also process more complex motor tasks that involve multiple sensory inputs and analysis. For example, Schlerf et al. found that these regions became activated when they asked their adult subjects to use their toes to write their signature. Simply zigzagging one's toes only activated the traditional (and phylogenetically older) anterior portions of the cerebellum. They concluded that one of the functions of the neocerebellum reflected higher cognitive levels of motor planning, rather than just simply processing sensory input (but it was still involved with the latter's processing).

Cognitive neuroscientist Kenji Doya (2000) and others (e.g., Caligiore et al., 2017; Moberget & Ivry, 2016) typified the cerebellum's functions well when they claimed it may be best understood for the neural computations it performs. In their model of cerebellar functioning, the cerebellum creates internal models of the environment for subsequent predictive control, and so it may be likened to a device that performs supervised learning, while the neocortex may perform unsupervised learning.

The Cerebellum, Basal Ganglia, and the Neocortex

There is also increasing evidence that the cerebellum and basal ganglia work in concert with the neocortex. The basal ganglia are a group of subcortical neurons with connections to the cerebellum, neocortex, thalamus, and brainstem. One group of basal ganglia nuclei (a cluster of neurons) is the substantia nigra, which is one of the principal producers of the chemical neurotransmitter dopamine (which heavily regulates intentional motor movements and socially and emotionally related behaviors). Parkinson's disease initially consists of a predilection for the destruction of cells in the substantia nigra, its symptoms being disruption of the ability to conduct voluntary motor movements because their execution depends on the striatum portion of the basal ganglia, and the striatum's performance is highly dependent on dopamine. The cerebellum also has reciprocal nerve tracts with the striatum (by way of the thalamus). The basal ganglia are involved in procedural learning and memory and in other learned or routine habits, and they appear to regulate the smooth and seamless execution of these tasks.

The basal ganglia have also been implicated in the selection of which actions to perform, in conjunction with the prefrontal cortex and cerebellum.

The Cerebellum and Classical Conditioning

It has recently been shown that the cerebellum is a key neurological component in classical conditioning. There are two common forms of classical conditioning, appetitive and aversive. The story of Pavlov's dog in Chapter 2 is an example of appetitive (desirous) conditioning, where the dog learns to respond to a new form of reward (the sound of a bell, which has been paired with food and elicits a consummatory response). In aversive conditioning, the organism learns to avoid the unconditioned stimulus (UCS) by responding (defensively) to the conditioned stimulus (CS) heralding it. It is important to note that both forms are critical to an organism's survival.

Appetitive conditioning has been studied less than aversive conditioning, although Martin-Soelch, Linthicum, and Ernst (2007) have shown that, in people and other animals, the amygdala, anterior cingulate cortex, orbitofrontal cortex, striatum, and other basal ganglia structures are all involved in the establishment of appetitive learning. While these brain structures do not directly implicate cerebellar function, most of them have intimate reciprocal relationships with the cerebellum. The authors note, "This process [appetitive conditioning] certainly involves more regions, such as the cerebellum, as evidenced in neuroimaging studies, but little is known about their specific role in appetitive associative learning" (p. 10).

In regard to aversive conditioning, Poulos and Thompson (2015) have demonstrated that the cerebellum is directly involved in the "learning" process of aversive classical conditioning, presumably in all vertebrates, including humans. Using rabbits, they showed that the cerebellum was specifically involved in learning not only to respond to a CS (a light or tone) that heralded the aversive UCS (or US; a puff of air to make an eyelid shut) but also the precise timing between the CS and the UCS. In their words, "the eyelid closure response (CR) becomes *very precisely timed* [italics mine] so that it is maximally closed about the time that the air puff or shock (US) onset occurs" (p. 253). They also concluded that this form of cerebellar learning was phylogenetically very old, as it has even been demonstrated in fish.

Cerebellar Processing: Forward and Inverse
Internal Models

Ito (1970, 1984, 1993, 2008) was one of the first neuroscientists to propose an "internal model" of processing for the cerebellum. According to Ito, the cerebellum has evolved to form and encode models of the dynamic action of body parts when interacting with their environment. This internal model for each body part is continually adjusted as movements are repeated. Through the highly reciprocal neural network of the frontal (especially the prefrontal cortex), temporal, and parietal lobes with the cerebellum (by way of the thalamus), the brain uses the continually adjusted internal model ultimately to perform the movement increasingly more precisely. Ito further contends that there will be less need to receive feedback from the moving body part, which may explain why movements become more and more skillful with practice. Ito claims that this internal model can also use the current sensory information to anticipate and predict future sensory states, which implies that this cerebro-cerebellar system learns to simulate relationships between commands for motor movements and their outcomes. The context-specific dynamics of the organism's interactions with its environment are conceptualized within Ito's internal model. Thus, just as an internal model of a motor movement could generate predictions about the sensory repercussions of a movement, an internal model could be generated for more abstract concepts, ideas, and thoughts. The latter is the basis for Ito's claim that the cerebellum may treat thoughts just like motor movements. Schmamann (1991) had similarly proposed that the cerebellum regulates the speed, consistency, and even the appropriateness of cognitive processes, and subsequent research provides strong support for this.

Ito's forward internal model makes a fundamental assumption that the cerebral cortex works explicitly (consciously) and the cerebellum works implicitly (unconsciously). An explicit thought forms a particular mental model consistent with internal goals and motivations. All three major regions of the prefrontal cortex (dorsolateral, ventromedial, and orbitofrontal) help form this explicit mental model with input from the temporal and parietal regions, the latter of which may actually instantiate the initial model given their visualization (episodic and otherwise) and visuospatial abilities. In Baddeley's working memory model, which does not specify brain regions, this explicit mental model would probably be formed in the episodic buffer. An implicit internal model is then formed by the cerebellum based on the explicit model, and it is not thought to be subject to conscious recognition or awareness, but a person might be aware that they are thinking about something. Differences (error signals) between the explicit and implicit models constructed in the brain help the cerebellum tweak the implicit model, and when satisfied (a feeling of intuitive reasonableness), that internal model is passed forward to the prefrontal cortex for their final execution.

There is no debate about the generation of internal models by the cerebellum. However, there is some debate about the nature of the models—that is, are they processed forward or inversely? The forward model posits that a movement command is transformed into a prediction of the sensory outcome of the movement's generation. Presumably, the inverse model inverts this same information flow. Also to be determined is whether only the phylogenetically newer parts of the cerebro-cerebellar network provide these forward or inverse models. In terms of just motor control (i.e., limb movements), it appears that a forward model has substantial empirical support, but there are some studies that favor inverse models. It is also possible that both models reside in the cerebellum but in different regions. The debate continues (Ishikawa, Tomatsu, Izawa, & Kakei, 2016; Koziol et al., 2014), but I have tried here to focus on the ultimate outcome of the models and the implications for the cerebro-cerebellar network rather than their specific processing intricacies. A broad consensus paper is also available which nicely summarizes the role of the cerebellum in both movement and cognition and the many issues still to be resolved (Koziol et al., 2014).

Novel Problem Solutions and Intuition

How might this cerebro-cerebellar system solve novel problems and how might intuition be accounted for? Ito (2008) proposed that when encountering a novel problem, a satisfying solution may not be initially found. The perceived mismatch between the generated model and the novel problem activates the "novelty system" of the brain (probably synonymous with the dopamine system; e.g., Schmajuk, Lam, & Gray, 1996), which is primarily located in part of the hippocampus and basal ganglia, and these regions in turn activate the attentional nature of Baddeley's central executive. It appears that all novel cognitive tasks engage the hippocampus, basal ganglia, and the amygdala. The consequence of this activation is that the central executive component then is more heavily engaged, attentionally and otherwise, in finding a solution. If a satisfying solution is found through the reciprocal and iterative actions of the cerebro-cerebellar network, the new model is fed forward to the central executive (prefrontal cortices). If a satisfactory solution is not found, the central executive's attention to the novel problem eventually fades. Ito (2008) proposed that the central executive continues to act on mental models "but intermittently and with a lesser degree of attention" (p. 308). In his view, the implicit processing of models continues within the cerebellum until a satisfying solution is reached. Thus, the process is iterative and based on the aforementioned differences in the models' signals. Because the process is implicit and not under the guidance or constraints of conscious awareness, Ito said that a correct answer seems to appear intuitively, but in reality it is a function of the cerebro-cerebellar interplay of explicit and implicit modeling, and particularly because of the continued implicit thinking and modeling of a solution by the cerebellum.

The Cerebellum and Autism

One of the strongest lines of evidence for the cerebellum's role in cognitive processing comes from the link between cerebellar disturbances and autism. *Autism spectrum disorder* (ASD) is currently defined as a multisystem disorder that affects the brain, immune system, gastrointestinal tract, and social and emotional functioning. ASD has two major clusters of behavioral symptoms: (1) social and emotional dysfunction, and (2) repetitive/stereotyped behavior. Cerebellar dysfunction has been consistently linked to ASD more than any other brain region. This evidence includes neuroimaging studies and a host of genetic studies. Interestingly, imaging studies have shown that the neocerebellum (lateral and posterior portions, which receive input from the prefrontal cortex) is most often implicated in ASD. One of the major social difficulties is a defect in theory of mind (the ability to estimate the thoughts and attitudes of other people). From the internal modeling perspective, it might be assumed that people with ASD are unable to form internal models that accurately reflect the thoughts of others. This might occur because explicit models can be formed by the prefrontal cortices and temporoparietal cortices, but the cerebellum's contribution of complementary implicit models, which contain error-checking mechanisms, cannot be generated.

Support for theory of mind deficits in people with ASD comes from a study by Van Overwalle, Baetens, Mariën, and Vandekerckhove (2014), who performed a meta-analysis of 350 neuroimaging studies of the cerebellum and social cognitive tasks. They found only marginal evidence for cerebellar involvement in most social cognitive domains, with profound exceptions. There was universally high cerebellar activation (1) in tasks where mentalizing was highly abstract, for example, determining when groups of others' behavior involves particular traits or characteristics (i.e., stereotyping groups); (2) when imagining past personal events or imagining or simulating future events; and (3) in tasks requiring imagining hypothetical events (i.e., counterfactualizing). Again, this finding supports the advantages of the implicit modeling and error correcting functions of the cerebellum in the cerebro-cerebellar network, and the psychopathological repercussions when this system is dysfunctional.

It is also important to reemphasize that ASD is not only a multisystem disorder but also exhibits a high degree of phenotypic and genotypic variability. Thus, Yuen et al.'s (2017) claim that the genetic predisposition for ASD "may be different for almost every individual" (p. 603) helps to establish that ASD may have various distinct subtypes, each with a distinct genotype. Therefore, linking some types of ASD to cerebellar dysfunction may have a sound basis, but linking all forms of ASD to cerebellar dysfunction may not be warranted.

The Cerebellar Cognitive Affective Syndrome

In neuropsychology, *dysmetria* (literally meaning "wrong length") is a specific kind of ataxia (loss of control of motor movements) in which a patient cannot properly carry out an intended motor movement, and the person tends to misjudge the target by overshooting it, undershooting it, or to have an improper velocity or rate of the action. Schmahmann (1991) proposed that there also exists a "dysmetria of thought." He later proposed (2004) a *cerebellar cognitive affective syndrome* (CCAS), with the following symptoms:

(1) Problems with executive functions such as poor planning, deficient abstract reasoning, working memory problems, trouble with multitasking and set-shifting, decreased verbal fluency

(2) Impaired visuospatial cognition, including disorganization and poor visuospatial memory

(3) Personality changes such as flattening or blunting of emotions, disinhibition, or inappropriate behaviors

(4) Language difficulties, including troubles with prosody (appropriate variations in the melody, tone, quality, and accents used in speaking), word-finding difficulty, and grammatical errors (not associated with poor environmental conditions).

He also concluded that the syndrome was associated with an overall lowering of intellectual functioning. It is important to note that Schmahmann characterized this syndrome on the basis of patients with strokes, tumors, atrophy, or infections of the cerebellum. One interesting area of future research could be a screening for CCAS in populations where cerebellar damage was not suspected, to determine whether CCAS can appear where the cerebellum appears to be normal. In other words, in people not suspected of having cerebellar damage (yet the cerebellum might be compromised at some anatomical level), is it still possible to have word fluency problems; deficient abstract reasoning; inappropriate, impulsive, or erratic behaviors; poor grammar and strange word choices; neologisms; or highly superficial/superlative wordings (e.g., *awesome, terrific, a perfect conversation, the best ever, bigly*) because of subtle cerebellar compromise?

The Cerebellum and Creativity

American neuroscientist Larry Vandervert and his colleagues (Vandervert, 2003, 2015, 2016; Vandervert, Schimpf, & Liu, 2007) have been strong proponents of the role of the cerebellum in creativity and innovation. They invoke both forward and inverse internal models to explain the origins of creativity, known as the *cerebellar-creativity hypothesis*. They propose that forward models (aka predictor models or explicit models) predict the outcome of a motor movement or thought, after which the prediction is confirmed by the sensory consequences of the outcome. After repetition or practice, an inverse model is formed on the basis of the predictor model such that further interactions assessing the success or failure of the motor action can be bypassed. The inverse model can then accomplish the task automatically (implicitly) and without any conscious effort. In this manner, the model allow the "establishment of automaticity" (Vandervert et al., 2007, p. 7). It is also presumed that multiple predictor models are generated by the cerebellum, and they can be combined to produce a large repertoire of images, thoughts, or ideas. Vandervert reasons that just as the cerebellum forms an internal model for movement and is then able to generalize that single trajectory into multiple trajectories (also in Ito's model), the cerebellum can apply this internal modeling capability to images, mental spaces, and ideas as well, which creates a newly formed internal "mental space" model. Further, the process would allow the blending of this internal "idea" model with other internal models. The cerebro-cerebellar network thus examines and compares previously learned models and newly formed models to create a solution to a novel problem, and that, in essence, is creativity. In this manner, Vandervert and his colleagues have essentially proposed that a hierarchy of models is created with varying levels of abstraction. When the central executive component of working memory is engaged in the evaluation and manipulation of these models, they might be experienced within the visuospatial sketchpad as mental images, reflections, or daydreaming. It is also assumed that the unconscious automaticity that is established by the cerebro-cerebellar network frees the central executive component from its attentional demands and actions, thus creating an environment for combined mental models or a new mental model to arise. The conscious result might be then experienced as "sudden insight" or "intuition."

Recently, Saggar et al. (2015), interested in the neural correlates of artistic creativity, tested the creativity-cerebellum hypothesis in a clever word-guessing game (Pictionary™) played on a tablet within an fMRI scanner. They found that words that were rated as having a higher "creative" content did indeed engage bilateral portions of the cerebellum. They speculated that implicit models of the words' representations created by the cerebellum somehow facilitated manipulations of the representations by "simulating and parallelizing the sketching of the given word in multiple ways. Such simulations, would in turn, allow participants to more efficiently draw the target word" (p. 7). What remained unclear, however, is how this efficiency would be translated into greater creativity. It does seem possible that, again, the efficiency or automaticity that develops as a function of the models in the cerebro-cerebellar network frees the central executive component of its primary attentional duties, which in turn allows an allocation of attention to other cognitive processes such as daydreaming.

Evidence for the Abstractive Abilities
of the Cerebellum

Balsters, Whelan, Robertson, and Ramnani (2013) designed an fMRI study to test whether the cerebellum only processes prefrontal cortical information when a model specifies actions (first-order rules) or whether it can process any model of prefrontal information, even if it is highly abstract (second-order rules) and independent of an action's properties. They found strong support for the hypothesis that the parts of the cerebellum that project to prefrontal cortices engage both first-order and second-order abstract rules, which implies that the cerebellum contributes to abstractive cognitive control independent of motor control. The results of their study strengthen the overall support for the hypothesis that a major exaptation of the cerebellum and cerebro-cerebellar network is its engagement with cognitive processes as elementary as word learning and as sophisticated as language. Overall, there seems to be little question of an evolutionary trajectory of not only an increase in the size of the cerebellum in the *Homo sapiens* lineage but also its increasingly important role in higher-level cognitive processing.

Summary

1. The cerebellum (meaning "little brain" or "lesser brain" in Latin) is estimated to contain about 80% of all of the brain's neurons. Its original adaptation appears to be the seamless and smooth execution of motor movements.
2. The dendritic branching of the cerebellum's Purkinje neurons is the most complex and extensive in the brain.
3. The lateral and posterior portions of modern humans' cerebellum have a more recent phylogenetic origin than that of other hominins.
4. An exaptation of the human cerebellum is the cognitive control and refinement of higher cognitive functions, including lower- and higher-level abstract thinking.
5. Numerous empirical studies link insight and creativity to the cerebellum.

8

The Hippocampus

Based on de Lorenzo's (2014) selfish-metabolism hypothesis (Chapter 2), even in the prebiotic world, an organism's ability to explore an environment (initially a watery, chemical environment) in the interest of its metabolism may have been more important than replication. As brains evolved to guide effective movements, remembering where to go and where not to go (and their associated evolutionary fitness consequences) was an essential part of guiding movements. Thus, it is not surprising that a structure in the brain came under selective pressure (more specifically, the genes that developed that structure) to navigate one's environment and to remember one's environment. That structure, whose primary adaptation was spatial navigation, was the hippocampus and its associated networks (see Fig. 8.1). Interestingly, one critical sense in the earliest evolution of navigation was olfaction, and close neuronal ties remain between the human olfactory bulbs and the hippocampus. As noted earlier, all mammals have a well-developed hippocampus compared to that in fish, reptiles, and birds, although the latter have homologous structures. Even the cells of the hippocampus have differentiated roles: Place cells become active and rearrange themselves in new environments, thus creating new and stable maps of those environments (Burgess, Becker, King, & O'Keefe, 2001; Burgess, Maguire, & O'Keefe, 2002). Other specialized cells, grid cells, are able to approximate distances, forming an additional neuronal basis for spatial navigation (Moser & Moser, 1998; Solstad, Boccara, Kropff, Moser, & Moser, 2008).

As noted in Chapter 2, there are two major memory systems, explicit (declarative, semantic) and implicit (procedural/motor, episodic). Although the distinctions may be somewhat blurred depending on the nature of the task (perceptual, motor, cognitive), it does appear that they involve different neural pathways for their initial encoding and storage. It also appears that the hippocampus' was adapted for the encoding and storage of implicit memories, such as visuospatial locations and the effective guidance of movements, as noted earlier. These procedural abilities, of course, required other brain structures and neural networks, particularly the cerebellum (Chapter 7), the

Evolutionary Neuropsychology. Frederick L. Coolidge, Oxford University Press (2020).
© Oxford University Press.
DOI: 10.1093/oso/9780190940942.001.0001

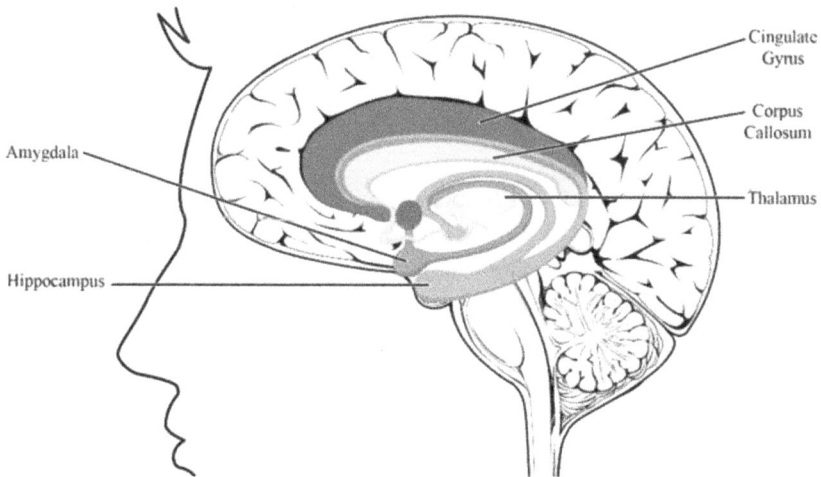

Fig. 8.1 Hippocampus and amygdala.
Source: Wikimedia Commons by Tara Dieringer.

frontal lobes (Chapter 4), and the parietal lobes (Chapter 5). However, it may be surmised that a brain structure and neural network came under major selective pressure to encode and store sounds and their meanings when primates diverged from the mammal lineage about 65 million years ago.

The major exaptation in the evolution of the hippocampus was neural reuse for the formation of declarative memories (also called *semantic memory, verbal memory,* or *explicit memory*). Prior to the development of more sophisticated neurophysiological measures of the brain, such as positron emission tomography (PET) scans and functional magnetic resonance imaging (fMRI), neuroscientists were dependent on EEG assessments and animal models, since human brain research usually involved brain-damaged patients, whose lesions were often haphazard because of the nature of brain disease or injury. EEG research has always been more effective in determining cortical (surface of the brain) functioning rather than subcortical functioning. Neuroscientists could virtually do anything they wanted (for good or ill) to study the hippocampus in rats and other animals, but because of EEG issues and artifacts (brain waves have a extremely low voltage and are subject to many sources of electrical interference) their suppositions were often dangerously flawed. Research on the hippocampus of animals was fruitful, but there was one important aspect of its function that could not be studied: declarative memory. And as I shall soon discuss, one of the most famous patients in neuroscience history, H.M., revealed to science the critical importance of the hippocampus to the formation and storage of declarative memories. Also, similar to the story of Phineas Gage, there were tragic consequences of his accident, which may have started as innocuously as falling off his bicycle.

As noted in Chapter 2, the hippocampus is considered part of a network called the *limbic system* (although inclusion of various brain regions in this network is a bit arbitrary), and the limbic system is well known for its role in memory formation, emotional processing, and fear. Further, as noted repeatedly, the connection between emotional states and memory is not inadvertent. Humans and many other animals remember things that have a strong emotional valence and tend to forget things that do not. The emotional valence of an event (which is highly dependent on one's prior experience, beliefs, or attitudes) will determine the strength of a memory. This phenomenon helps to explain the development of posttraumatic stress disorder (PTSD) in some humans. The evolutionary origins of PTSD will be discussed in Chapter 10.

Other Structures of the Medial Temporal Lobes

The hippocampus is surrounded by the parahippocampal gyrus (approximately BA 27, 28, 34, 36, 37), and it also considered part of the limbic system. The anterior part of the parahippocampal gyrus includes the entorhinal and perirhinal cortices. One region of the parahippocampal gyrus, the parahippocampal place area, plays a recognition role for places, much like the adjacent fusiform gyrus does for faces and objects. The fusiform gyrus lies just below the parahippocampal place area and just above the inferior temporal gyrus. The parahippocampal place area appears to encode (converting into a form useable by neurons) and recognize environmental landscapes and scenes, thus it plays an important role in episodic memory. If the parahippocampal place area is damaged but not the fusiform gyrus (an admittedly rare event), a brain-damaged patient will be able to recognize the individual faces and objects in the landscape or scene but will not be able to visually recognize the landscape or scene.

The entorhinal cortex (BA 28 and 34) sits at the anterior end of the parahippocampal gyrus, and its lateral and medial portions appear to be dedicated to different functions (see the following discussion). The entorhinal

cortex has strong reciprocal connections to all cortical lobes and to the hippocampus, thus it serves as the major interface between all of the cortical
lobes and the hippocampus. It also forms part of an extensive network for
encoding, processing, optimizing, and recalling declarative/semantic, episodic, and autobiographical memories, as well as aiding spatial navigation.
The medial portion of the entorhinal cortex is where neural maps of the external environment are formed by grid cells, for which the neuroscientists
Edvard and May-Britt Moser received the Nobel Prize (as noted in
Chapter 3). The medial regions of the entorhinal cortex also estimate the
speed of an animal's movement and help guide directional movements from
proprioception (information from the senses like joints, muscles, and con-
nective tissues, and body position and movement) based on the firing rates
of the grid cells. It is thought that the hippocampus then processes that general information into unique representations of the animal's environment.
The lateral neurons of the entorhinal cortex show little spatial selectivity but
are thought to represent non-spatial information which is eventually conjoined with spatial representations and streamed to other subcortical regions
(Hargreaves, Rao, Lee, & Knierim, 2005).

A recent study by Danjo, Toyoizumi, and Fujisawa (2018) demonstrated
the allocentric (from another's perspective) and conspecific (recognizing
one's own species) nature of hippocampal cells. Specifically, they found that a
particular set of cells in the hippocampus had spatially receptive fields for the
recognition of one's self and others, even when self or others are engaged in
movements. This ability to be aware of another's position in space is critically
important to all social animals, especially primates, in order to be able to
learn by observation, to engage appropriately, and to be successful in social
interactions and navigation while in groups. I would also speculate that this
conspecificity of hippocampal cells may be age-attuned, as human infants
and children will often pay greater attention to others their own age before
elders.

Of Moles and Men: The Evolution of Olfactory Bulbs in *Homo sapiens*

The olfactory bulbs (bilateral structures that look like the length and end of
a Q-Tip®) and olfactory cortex are also considered an important part of the
limbic network. Paleoneurologist Marcus Bastir and his colleagues (2011)

claim that the olfactory bulbs were greater in size in *Homo sapiens* than in Neandertals, based on the absolute and relative size of their cribriform plates (through which the olfactory receptors pass their sensory information into the olfactory bulbs, just above the cribriform plates). Thus, they reasoned that the larger the cribiform plate, the larger the olfactory bulbs.

The subtitle of this section is a pun, as moles (a subgroup of mammals, the insectivores) have an interesting commonality with *Homo sapiens* and not with Neandertals. Moles and modern *Homo sapiens* share reduced occipital lobes (i.e., visual cortices) relative to other parts of the brain, and enlarged olfactory bulbs. Although often overlooked, olfaction has played a critical role in evolution, and it continues to play an important role in reproductive success and immune system functioning. As human olfactory bulb size has been shown to be correlated to olfactory performance, that is, odor threshold detection, it is possible that there may have been some evolutionary advantages to a heightened sense of smell in *Homo sapiens* that Neandertals did not possess.

It has been demonstrated that olfactory neural circuitry is highly and intimately integrated into cerebral regions involved in higher cognitive functions, particularly the limbic system, where memory and emotions are intertwined, and it projects to brain regions (orbitofrontal cortex, hippocampus, amygdala) involved in mating, fear, reward, motivation, and stimuli evaluation systems (i.e., saliency or good vs. bad). Thus, it can be argued that olfaction is a higher cognitive function and serves a much more sophisticated role than simply smell. Animal studies of olfactory bulbectomies have shown such damage to be a causal factor in immune reactions and disease immunity. Human olfactory impairments are associated with autoimmune diseases such as lupus. Human studies have also shown that olfactory impairments herald many neurodegenerative diseases, including Parkinson's and Alzheimer's disease. Olfactory impairments have also been demonstrated in diseases known to have significant social and emotional impairments, such a schizophrenia and obsessive-compulsive disorders. Interestingly, congenitally anosmic (lack of smell) people appear to have significantly more social and reproductive problems than congenitally blind or deaf people (although this has not been empirically well substantiated). Two possible hypotheses for these complicated relationships focus on olfactory ensheathing cells, which may have differential regenerative capabilities and may be involved in olfaction's relationship to inflammatory processes, which have been shown to be one of the causes of neurodegenerative diseases. Thus, even marginally larger olfactory bulbs and better discriminatory power of olfactory receptors in *Homo sapiens* may have resulted in significant behavioral consequences regarding general immunity, resistance to autoimmune diseases, and social, emotional, and sexual/mating functions.

Imagine also the evolutionary role of olfaction in toxin detection. Touching a toxin is a poor way of determining its lethality, which is true for visual (it looks bad) or auditory (it is making bubbling sounds) discriminations. And tasting a toxin to determine its lethality could be, well, lethal. Among the senses, olfaction would likely be the safest means of determining whether a substance is harmful or not. Interesting also is the fact that adult brain neurogenesis (creation of new neurons in adults) is largely confined to just two regions: the hippocampus and the olfactory bulbs. Perhaps this is because of their more critical role in survival than other structures. Maintaining the ability to move and navigate successfully and to detect odors, not only of toxins but also of chemicals like pheromones (chemicals some animals release into the environment which affect the behavior of others), must have played a significant role in the evolution of the human species, and those abilities continue to play a significant role in the lives of modern humans.

Olfactory Receptor Evolution

Given olfaction's importance in evolution, it is not surprising that olfactory receptors are coded for and controlled by the largest family of genes in the genomes of all mammals. Interestingly, however, gene studies have shown that in modern *Homo sapiens* over half (~55%) of the olfactory receptor genes have lost their function. This loss of gene function usually occurs as an adaptation to a particular ecological niche, for example, the genes coding for temperature regulation are different in the tropics or cold regions. In a study of olfactory receptor genes in modern *Homo sapiens*, Neandertals, and Denisovans, researchers have found that the latter two groups showed an additional loss of olfactory gene function suspected to be due to their adaptation to colder environments, where cold temperatures reduce the effect of odors (Hughes, Teeling, & Higgins, 2014). This finding tenuously supports Bastir et al.'s (2011) finding that olfactory performance may have been compromised in the Neandertal lineage.

Amygdala

The amygdalae (plural form) are almond-shaped structures (its name is derived from the Greek word for "almond") that sit bilaterally at the anterior tips of the hippocampus. Many specialized amygdalae nuclei (bundles or clusters of neurons) have been discovered (e.g., lateral, basal, accessory basal, central, medial, cortical), so it is difficult to describe a common function to the entire amygdala. Despite the complexities of the amygdala's behavioral functions, it has long been well known for its role in fear conditioning, emotional processing, sexual behavior, and hunger (e.g., Delgado, Nearing, LeDoux, & Phelps, 2008; LeDoux, 1996). It is important to note, however, that an overwhelming majority of the studies of the amygdala only deal with fear.

As fear would be absolutely critical in the evolution of animals, the amygdala is fully functional at birth in humans, and it appears to be more reactive in childhood and adolescence than in adulthood (Tottenham & Sheridan, 2010), which makes sense, because children and young adolescents are physically more vulnerable. The amygdala is also well connected to other parts of the brain, with major direct connections to the hippocampus, all of the senses (through the thalamus), the ventromedial prefrontal cortex, and the olfactory cortex. These regions collectively form a complex neural network for processing the saliency of emotional stimuli, including the recognition and regulation of both positive and negative emotions.

Recognition of the Critical Functions of the Hippocampus: The Story of H.M.

It was September 1, 1953. A 27-year-old man, subsequently known only by his initials, H.M. (until his death in 2008), was about to undergo a bilateral hippocampectomy and a bilateral amygdalectomy (the removal of brain tissue from both hemispheres). H.M. had been knocked unconscious in a bicycle accident when he was 9 years old, although neurological testing and a physical examination at that time were normal. Interestingly, there were no witnesses to his accident, and because he at the very least suffered a severe concussion, which usually is accompanied by vague reports from victims, it is not certain whether the bicycle accident was actually the cause of his head injury. At the age of 10, he began to have minor seizures in which he would "blank out" for about 40 seconds and be unresponsive. At 16 years old, H.M. started having major seizures. He would bite his tongue, lose consciousness, become incontinent, and fall into a deep sleep afterward. His EEG results did not reveal any single area as a possible cause of his seizures. Although a majority of major epileptic seizures can be controlled by medication, H.M.'s seizures were not, and they became more severe and more frequent, to the point where H.M. could no longer work at his job as a motor coil winder. Thus, in 1953, neurosurgeon William Scoville attempted to relieve H.M.'s seizures by removing his hippocampus and most of his amygdalae.

Previously, Scoville had performed numerous psychosurgeries, including frontal lobotomies, primarily in an attempt to improve the psychological functioning of seriously disturbed mental patients (schizophrenic patients). Along with neurosurgeon Walter Freeman II, they had experimented with various forms of lobotomies in these patients. The lobotomies were performed without much success in changing the patients' personalities, although one serious side effect eventually became well known, severe apathy. As a result, the popularity of the operation declined. With the exception of severe apathy after a lobotomy, it was noted that the schizophrenic patients rarely got worse in terms of their psychotic thought processes. Moreover, the lobotomy procedure did not appear to affect any of the patients' intellectual reasoning or intelligence. In truth, however, the lobotomized patients were rarely, if ever, tested afterward for psychological or cognitive problems.

Scoville did not like Freeman's particular lobotomy method, known as a "ice pick" lobotomy (see Fig. 3.5) because he used a thin stainless steel instrument about 30 cm long (12 in.). Scoville apparently found that method sloppy, so he began experimenting with surgically ablating (removing) the medial and posterior regions of the frontal lobes and portions of the temporal lobes, where the hippocampus and amygdalae reside medially. Primarily, he wanted to see if there were any therapeutic effects to be gained by ablating the temporal lobes instead of the frontal lobes. At this time, it was also an accepted medical procedure to remove parts of the temporal lobes in order to reduce the severity of epileptic seizures or to stop them completely, but only if the cause of the seizures could be identified as a single epileptogenic focus (the area causing the seizures). Further, in other epileptic patients, the temporal lobes, hippocampus, and amygdalae were the most frequent epileptogenic areas. Thus, it seemed reasonable to Scoville that a bilateral removal of the hippocampal area (including most of the amygdalae) might reduce

H.M.'s seizures and allow him to return to at least a semi-normal life, rather than one continually interrupted by devastating seizures. During the operation, and before any tissue was removed, the medial surfaces of H.M.'s temporal lobes were exposed and tested for epileptogenic qualities. Again, there was no single area found to be responsible for the seizures, but they were removed nonetheless.

After the operation, H.M. appeared to be a normally functioning adult with fully intact understanding, speech, and reasoning (i.e., neurotypical). He was described as quiet and polite. There were no apparent "hard" neurological deficits like paralysis, or any visual or perceptual problems. However, H.M. no longer recognized the names or faces of the hospital staff as he had done so easily before the operation. He could not find his way to the bathroom. He also could not remember events dating to about 3 years before his operation, but his earlier life memories were clear and intact (called *partial retrograde amnesia*). About 2½ years later, H.M. was examined by clinical neuropsychologist Brenda Milner (and others), who specialized in the evaluation of personality, intellectual functioning, and memory as a result of brain damage or brain disease. About a month before his operation, H.M.'s IQ was 104 (just slightly above average). During Milner's examination, his IQ was 112. The increase in IQ was attributed to the fact that H.M. was no longer having severe, daily seizures. Unfortunately, H.M. still had severe seizures (called *grand mal*), but the frequency was greatly reduced. Before the operation, he also had suffered minor seizures (petit mal or absence seizures) as often as every 10 minutes throughout the day, but after the operation his minor seizures were down to only one or two daily.

Despite his now slightly above-average IQ, his verbal memory tests were at least two standard deviations below average. His memory was most severely interrupted in the area of new verbal learning (*anterograde amnesia*). He had "zero" scores for learning difficult new words and low scores for learning even easy words, and his performance did not improve with practice. Further, H.M.'s mother corroborated these findings outside the testing places. She said they had moved to a new home about 2 years after his operation. In the 10 months since the move, H.M. could not remember the new address, nor could he find his way to the new house. However, he could remember his old address perfectly. He could not find the lawn mower a day after he had used it. He would read the same magazines over and over without remembering that he had already read them. He would work and solve the same jigsaw puzzle, yet not show any improvement from practice. It was later reported (H.M. lived until he was 82 years old, 55 years after his operation) that he liked crossword puzzles. However, closer examination revealed that he often started them but rarely completed them. He would also eat lunch and not be able to describe a single thing he had eaten, even 30 minutes later. Interestingly, H.M. never complained about being hungry or having headaches or stomachaches. He would eat if the food was placed in front of him and would eat normally even if the food had been delayed, but there were no complaints about being hungry (an interesting trait, given the role of the amygdala in promoting and suppressing hunger). Also, during Milner's evaluation (April 26, 1955), he gave the date as March, 1953 (a date that was 6 months before his operation) and he said he was 27 years old (when he was about 30 years old). Later in his life, H.M. was to remark, "Every day is alone in itself, whatever enjoyment I've had and whatever sorrow I've had." And, unfortunately, his doctors had the impression that most of his experiences during the day faded long before the day was over. He often described his own mental state "like waking from a dream."

Milner and her colleagues re-evaluated H.M. in 1962 and again in 1966, about 14 years after his operation. His mother reported (and it was confirmed through testing) that his anterograde amnesia was still severe, although he had shown some recent improvement. He still could not remember people or their names (e.g., neighbors, friends, or hospital staff) whom he met after the operation. He could always remember his birth date, events while growing up, high school experiences, and jobs he held in his 20s. However, he could not remember the current date and could only give wild guesses, sometimes using the outside weather as a clue. Also, he still had difficulty remembering events that had occurred up to about 2 or 3 years before the operation.

In his evaluations in 1966, his IQ continued to improve (118), but his verbal memory was still devastatingly low. His mother reported that he still liked puzzles, although he would solve the same one over and over without remembering that he had done so. He also was unimpaired on a block design test, in which different-patterned red and white blocks had to be assembled within time limits. His ability to repeat digits forward and backward (i.e., phonological storage capacity) seemed to be unimpaired as well, although it was perhaps less than expected given his now well above-average IQ. It is also important to note that his mother's reports were frequently used because H.M.'s parents housed and cared for him after the operation, since he was incapable of living independently.

A Surprising Memory Finding

One of the most dramatic surprises came when H.M. was tested for his procedural memory using the reverse mirror-drawing task, which requires a patient to draw within the outline of a star by watching their hand in a mirror. H.M. was able to learn this task, perform it normally, and retain this performance ability on subsequent testing. However, on subsequent testing, he could not remember whether he had done the task before. He could only demonstrate that he remembered the task by his performance.

It is also interesting that H.M.'s neuropsychologists noted that he seemed impervious to the pain of electric shocks (why and how they tested him with electric shock is a bit disconcerting), he was very even-tempered, and he lacked sexual interests. Current thinking, particularly concerning research on the amygdala in people and animals, is that it may play a profound role in aggression, sexuality, eating disorders, and emotion. Some research supports the hypothesis that electrical stimulation of the amygdala may result in aggression, hypersexuality, binge eating, obesity, and inappropriate emotions. Removal of the amygdalae, least in animal models, may result in passivity, lack of sexual interest or drive, lack of appetite, indifference, and sometimes even-temperedness.

In summary, H.M.'s bilateral hippocampectomy and amygdalectomy had some subtle and some profoundly devastating effects. When I was in graduate school in the early 1970s, I had a neurophysiology professor who had written that "an impairment of recent memory has not been a distinguishing characteristic of hippocampal damage" (Isaacson, 1972, p. 53). So it appears that the scientific world at that time, even nearly two decades after H.M.'s operation, was still slow to recognize that the hippocampus did play a critical role in establishing declarative (semantic) memories. Part of the problem, of course, was that animals can be tested for procedural and visuospatial (maze running) memories but most animals (like rats) cannot be trained or tested for declarative or semantic learning and memory. Also, it is important to remember that on a casual meeting, H.M. would appear virtually normal. There was nothing about his speech, reasoning, or perceptual abilities that would make someone take notice. He was polite and would apologize if he had forgotten a person's name during a conversation. H.M. walked cautiously with a "broad-based gait," but his examiners decided it may have been a pre-existing condition, which may have resulted from spinal cord damage (perhaps from his bicycle accident). Yet, the unfortunate nature of his devastating cognitive impairments was that 14 years after his operation, he needed to live with and be cared for by his parents, and he could not even find his way home. He did have a job, but it consisted of placing cigarette lighters on a display, because he labored in a sheltered workshop for cognitively impaired adults. Important to note also is that his IQ at that time was higher than about 85% of the general population.

Did removal of H.M.'s amygdalae account for his meekness, even-temperedness, lack of hunger, and his lack of interest in sex? These behaviors are more difficult to speculate on, since they are more difficult to test, and it is difficult to assess H.M.'s premorbid (before his operation) behavior. However, some scientists (e.g., Corkin, 2013; Dittrich, 2017) suspect that he was this way before his operation, and his current behavior may not have reflected a dramatic change. However, given the current body of empirical research on the amygdala, it does remain likely that H.M.'s behavior in these areas may, at the very least, have been maintained by the absence of his amygdalae. Interestingly, MRI scans and a brain autopsy performed after his death, in 2008 (Augustinack et al., 2014), revealed that extended portions of his amygdalae had survived. Further, it was determined during his autopsy that his olfactory bulb and tracts were completely intact; however, his olfactory cortex (BA 28 and 34) had been removed during his operation, so H.M. was completely anosmic.

A Summary of Functions of the Hippocampus

What did neuroscientists finally learn from H.M.'s experience? First, the hippocampus plays a critical role in the formation of verbal memories. Second, it plays a critical role in the transfer of newly acquired verbal memories into long-term memory storage. It is now known that the memories themselves are not stored in the hippocampus itself but throughout the cerebral cortex. Third, it appears that this transfer does not occur all of a sudden, as H.M.'s memories of 1 to 3 years ago still remained vulnerable to forgetting, thus any severe form of physical disruption to the brain, like H.M.'s psychosurgery, blows to the head, electroshock, or even aging, may negatively affect the transfer of memory to long-term storage. Was H.M. completely unable to learn new declarative memories? No, although it took an intensive amount of practice, and even then this memory was spotty. Apparently, he was able to learn the immediate surrounding neighborhood after his parents had moved, although this also took him a vast amount of practice. Furthermore, when asked by his doctors to direct them to his house, he directed them to his old neighborhood. Thus, he was not completely unable to learn new declarative memories, but it took an inordinate amount of practice and the quality of the new memory was compromised. Fourth, neuroscientists learned that procedural memories are probably not dependent on the hippocampus for their formation or storage, since H.M. was able to learn the reverse mirror drawing task. However, H.M.'s ability to verbally state that he remembered the task was impaired; thus, even a simple task, like mirror drawing, appears to have both a procedural and declarative memory component. Also, H.M. did not perseverate (make the same mistake over and over) on frontal lobe tasks; thus, they knew that hippocampal damage was qualitatively different than frontal lobe damage. This finding was also true for the facial recognition memory task, on which H.M. performed poorly and frontal lobe–damaged patients are unimpaired. Fifth, it became abundantly clear that there were at least two powerful but different memory systems operating in humans, declarative (explicit) memory and procedural (implicit) memory, and that there were different neurological bases for each of these systems. Further, the original adaptive function of the hippocampus was maintained in humans: visuospatial abilities and place recognition. The memory for smells was also preserved in humans; its role in human evolution was previously discussed. However, the cells of the hippocampus were exapted for the processing of verbal and acoustic memories, as reliance on these systems became more profound with the emergence of primates.

Memory Interactions involving the Hippocampus, Cortices, and Sleep

Because H.M.'s basic phonological storage system was intact (according to his performance on the digits forward and backward task), it is reasonable to assume that he could maintain newly heard or seen information in his cerebral cortices but, of course, only temporarily. This implies that both the cortex and the hippocampus have a role in initial learning, but what follows (i.e., hippocampal integration) leads to memory consolidation and long-term storage. In a recent study, Genzel and Battaglia (2017) found that an extended period of sleep (greater than 2 hours) following the replay of a recent episodic memory leads to memory consolidation in the cortex. However, recent memories followed by additional stimulation (novel environments or novel perceptions) strengthens hippocampal memory formation and consolidation. Further, they found that memories consolidated during sleep are less prone to interference and more stable than memories followed by additional stimulation or learning. Overall, these findings reinforce the general hypothesis that sleep plays an important function in the consolidation of memory, and, indeed, as discussed more fully in Chapter 9, sleep may be an ideal state for memory consolidation and enhancement. In addition, it appears that there is a dynamic interplay between the hippocampus and cortical regions, which also interacts with sleep states in a highly beneficial fashion.

Summary

1. All mammals have a well-developed hippocampus compared to that in fish, reptiles, and birds, although the latter have homologous structures also involved in spatial cognition.
2. The cells of the hippocampus have differentiated roles: Place cells become active and rearrange themselves in new environments, which create new and stable maps of those environments. Grid cells are able to approximate distances, forming an additional neuronal basis for spatial navigation.
3. The hippocampus and olfactory bulbs have intimately related functions.
4. The story of patient H.M. revealed that declarative memories are consolidated by the hippocampus, but procedural memories can be established without hippocampal involvement.
5. Declarative memories remain vulnerable to disruption and forgetting up to about 3 years after memorization.
6. Memories consolidated during sleep are less prone to interference and more stable than memories followed by additional stimulation or learning, although the latter creates stronger memory traces.

9

The Evolution of Sleep and Dreams

As sleep researcher Allen Rechtschaffen (1971) has noted, "If sleep does not serve an absolutely vital function, then it is the biggest mistake the evolutionary process has ever made" (p. 88). But of course, evolution did not make a mistake, although there are many evolutionary dead ends. While the vital purpose of the evolution of sleep may not be completely settled, it is safe to state that sleep and dreaming evolved for many reasons. It also appears that there have been many different exaptations of this state of "quietude" throughout its evolution.

Returning yet again to earth's origin, all life developed in a 12-hour (on average), light–dark cycle (aka circadian rhythm) because the earth rotates on its axis completely every 24 hours (actually every 23 hours and 56 minutes) as it revolves around the sun. If life had instead evolved on Mars, these circadian rhythms would likely be similar since Mars rotates on its axis every 24 hours and 37 minutes. It would be vastly different had life evolved on Venus, which rotates completely every 243 earth days! *Sleep* can be defined as periods of quiescence, immobility, and a relative lack of intentional contact with one's environment, but it also includes responses to internal stimuli. If *wakefulness* is defined as periods of activity, mobility, and responses to internal and external stimuli, then virtually all animals exhibit a sleep–wake circadian rhythm. Further, virtually all living organisms, including animals, plants, fungi, and even bacteria, have adapted their physiological processes to this circadian cycle. However, it cannot be concluded that all living organisms actually sleep, as some bacteria may be said to be in a constant state of varying activity.

Primitive sleep states undoubtedly developed from the natural light–dark cycle, which constrained activities to periods of light and inactivity to dark periods. Primitive sleep states were also probably constrained by seasons, where activity was more likely during warm seasons and inactivity (i.e., hibernation) during cold seasons. Of course, as soon as such environmental niches (suitable places) were established, some organisms reversed these patterns in order to improve or avoid predation. Perhaps primitive sleep

Evolutionary Neuropsychology. Frederick L. Coolidge, Oxford University Press (2020).
© Oxford University Press.
DOI: 10.1093/oso/9780190940942.001.0001

states can be traced back to the first swimming metazoans, simple flatworms, appearing about 545 million years ago. Studies with roundworms like *Caenorhabditis elegans* show that they have resting states akin to sleep during molting (Raizen et al., 2008), whereas studies with fruit flies (*Drosophila melanogaster*) have shown alternating activity–rest periods (Cirelli & Bushey, 2008). During their "quiet" periods, fruit flies are more difficult to arouse, and just as in mammals, there is individual variability in their sleep–wake–rest cycles (chronotypes), with consistent variation in gene expression being the driving force for metabolic activities and chronotypes. As fruit flies are descendants of insects (land-bound crustaceans), which date back at least 400 million years, it can be assumed that many of the fundamental features of modern human sleep were in place between 545 million and 400 million years ago.

Modern reptiles, which began evolving about 300 million years ago, also have sleep–wake cycles. Their brains exhibit a deep-sleep state similar but not identical to the slow-wave sleep (SWS) of mammals and primates. The similarity is the high-arousal threshold during this state of sleep; however, modern reptiles do not show the slow EEG waves of about 1 to 3 Hertz (Hz), nor the large amplitude associated with modern SWS in humans. Every extant bird species shows SWS and rapid eye movement (REM) sleep, and because birds evolved from dinosaur-like reptiles about 250 million years ago, these two fundamental sleep characteristics were in place even before the evolution of mammals, about 200 million years ago. All mammals have well-defined SWS and REM sleep, although it is possible that the evolution of these two sleep stages evolved independently of the deep sleep of reptiles.

As noted in Chapter 3, mammals have an enlarged cortex and hippocampus compared to reptiles, so it is possible that the evolution of SWS and REM sleep was due to their more sophisticated cerebral cortex, with its thicker, six-layered surface compared to the thinner, three-layered surface of reptilian brains. Evolutionary biologist J. Lee Kavanau (2002) has proposed that the evolution of SWS and REM sleep accompanied the evolution of warm-bloodedness, which arose in mammals. His hypothesis relies heavily on the importance of organisms to learn and store new information (memory), maintain memories, and reinforce links between old and new memories. As noted earlier, this new and expanded cortex in mammals, the neocortex, gave them much greater behavioral flexibility compared to the more reflexive reptiles, although both animals required physiological systems that supported learning and sustained memories. That these learning and memory support systems co-evolved in response to light–dark cycles is not at all surprising.

It may be surmised that cyclicity developed between wakefulness and primitive sleep (rest or inactivity, but not necessarily SWS or REM). The environmental conditions that impelled the evolution of wakefulness required sustained attention, sensory and perceptual processing, new learning, and the appropriate recall of memories to solve novel problems—in other words, highly efficient brain functioning. The conditions of rest or inactivity may have evolved initially as an adaptation that conserved metabolic energy and acted as a defense mechanism against the dangers of nocturnal activity. The evolutionary value of this primitive sleep state, which itself may have made an animal more vulnerable to predation, had to have exceeded the fitness costs (liabilities) of inactivity. Thus, the exaptation of this primitive resting state to its use in the processing of newly learned information, its storage, and its integration with previous memories helped the organism overcome the vulnerability associated with a resting state.

Kavanau (2002) has further proposed that REM sleep, which is accompanied by fast cortical brain waves (i.e., >15 Hz) and voluntary muscle paralysis (muscle atonia), may have evolved in warmed-blooded animals when muscle contractions were not needed for thermoregulation. This occurs when ambient temperatures are approximately equal to an animal's core temperature, for example, at twilight and dawn. Further according to Kavanau, fast cortical waves, which also accompany wakefulness, help to bind and reinforce memories in a "coordinated" fashion. The waves in SWS reinforce memories in an "uncoordinated" way and in a way that suppresses coherent mentation; any recall of unsynchronized and unreinforced thoughts is essentially meaningless. This may be observed in human sleep subjects, who, when awakened from SWS, invariably deny having had dreams or coherent thoughts.

So, how did dreaming arise from REM sleep? Kavanau hypothesizes that the uncoordinated reinforcement of memory circuits during SWS without accompanying mentation was followed by the coordinated reinforcement of memories during REM sleep, which was now associated with unconscious awareness, or dreaming. Finally, modern human sleep patterns are characterized by SWS followed by REM, and as the sleep period progresses, REM sleep will be interspersed by periods of non-REM sleep. Thus Kavanau thought that the sequence of four or five cycles of non-REM and REM sleep must have had further cognitive benefits associated with uncoordinated reinforcement of memories followed by coordinated reinforcement of memories.

Sleep and Its Stages

Sleep Stages

There are currently two systems for the classification of sleep stages. The older system consisted of a restful yet awake stage (stage 0), four non-REM stages (stages 1 through 4), and REM (Rechtschaffen & Kales, 1968). The newer system collapses the four non-REM stages into three by combining stages 3 and 4 into non-REM 3 stage (see American Academy of Sleep Medicine [AASM]; Silber et al., 2007, for additional details). In the following discussion, I shall amalgamate the two systems (with AASM designations in parentheses). See Fig. 9.1 for sleep brain wave examples.

Stage 0 (W, or Wakefulness)
Humans and many other animals start their sleep period by resting with their eyes closed. On an EEG, a relatively high-amplitude and semi-symmetrical wave (alpha) appears, and these waves vary among humans from about 8 to 12

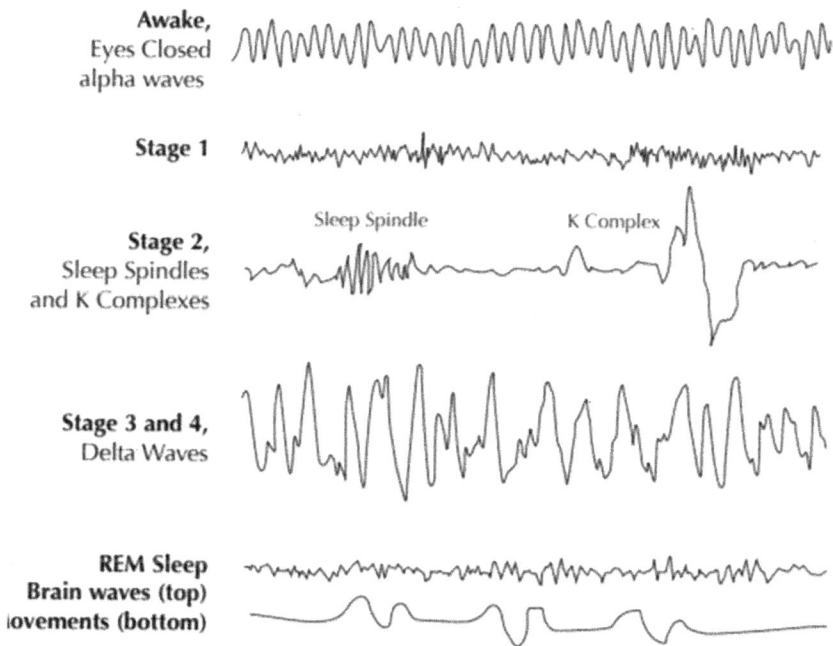

Fig. 9.1 Sleep stage waves.
Source: Holly R. Fischer, MFA.

Hz (but within a single person, the wave frequency will be consistent). Alpha waves were first described in 1929 by Hans Berger, a German psychiatrist, who is also credited with inventing the EEG. Alpha waves are measured by electrodes placed on one's head or skull, and they are the easiest to view when the electrodes are placed over the occipital lobes (primary visual cortex). Because alpha is usually detected when a person is resting or daydreaming, alpha has come to be associated with restful states of consciousness like meditation, thus its designation stage 0. Interestingly, about one-third of people do not produce alpha waves while resting with eyes closed, which makes claims that the alpha wave is indicative of a resting state specious. Also dubious are claims that inducing alpha states through biofeedback is akin to creating restful states. There are no good explanations for why some people do not naturally produce alpha waves, but those people deemed "alpha dominant" are preferred for sleep studies because it can be more easily determined when they actually enter a sleeping state. Most alpha-dominant humans will have some bursts of alpha throughout the night of a few seconds or more, which indicates they have briefly awakened. As people age, these periods of wakefulness between stages of sleep become more frequent.

Stage 1 (N1)

Stage 1 is usually the first and lightest (easiest from which to awaken) of the sleep stages. It accounts for about 10 to 15% of total normal human sleep. Stage 1 consists of theta (4 to 8 Hz), beta (12 to 30 Hz), and gamma (25 to 100 Hz) waves, which are usually of low amplitude, irregular, and desynchronized. All of these waves also occur during wakefulness, which is one of the reasons stage 1 is considered a transition stage from wakefulness to sleep. One interesting phenomenon during the onset of sleep in nearly all primates is the hypnic or hypnagogic jerk, which is a sudden reflexive muscle movement that frequently awakens the sleeper. Although the ultimate cause of the hypnic jerk is unknown, a common hypothesis is that it is an archaic reflex to the brain's misinterpreting the muscle relaxation accompanying the onset of sleep as a signal that the sleeping primate "thinks" it is falling out of a tree. The reflex may also have had the fitness benefit of having sleepers readjust or review their sleeping position in a nest or on a branch, thus ensuring against a fall (Coolidge, 2006).

Stage 2 (N2)

Stage 2 is also considered a light stage of sleep, with waves similar to stage 1. It is also the most prevalent stage of sleep, accounting for about 50 to 55%

of total human sleep. Two additional waves are characteristic of stage 2, sleep spindles and K-complexes. Sleep spindles have a slightly higher frequency (13 to 16 Hz) than alpha waves (8 to 12 Hz), and they appear once or twice every minute in bursts of 1 to 2 seconds. The other distinctive stage 2 wave is a K-complex, which consists of a single large, negative deflection followed by smaller positive ones. K-complexes can occur during wakefulness, but usually only in response to some external stimuli, and sometimes external stimuli during stage 2 sleep also elicit sleep spindles. Spindles appear to originate from the thalamus, whereas K-complexes appear to arise as an interaction between the cortex and thalamus. Is it suspected that they both play a role in memory consolidation and in keeping a person asleep.

Stage 3 and Stage 4 (N3)

Stage 3 and stage 4 have been combined in the AAMS system, and this combined stage is synonymous with SWS. It is called SWS because of the appearance of delta waves (0.5 to 3 Hz), which are the slowest frequency of all brain waves and have the highest amplitude. Delta-wave activity dominates this stage (greater than 50% of the brain waves during a 1-minute period will be delta waves). The SWS stage accounts for about 15% of total human sleep, and as noted previously, it has been observed in all mammals, birds, and in some fishes and reptiles. In humans, a majority of SWS occurs during the first third of the sleep period. SWS is also considered the deepest stage of sleep because it is during this stage that all animals exhibiting it have the highest arousal threshold. Interestingly, however, porpoises exhibit SWS in only one hemisphere while the other hemisphere is awake. In humans, studies have consistently shown that when awakened during SWS, nearly all sleepers deny having had any dreams, vivid or otherwise.

During SWS, the parasympathetic system (part of the autonomic nervous system, which controls vegetative functions like digestion) dominates, and the gastrointestinal system becomes more active. The cardiovascular system slows, with decreases in heart rate, blood pressure, and respiration. In humans, the percentage of SWS decreases across the lifespan, and the decrease is greater in men than in women (Moser et al., 2009). Whether this decrease in SWS is associated with variations in cognition, learning, or memory is not known.

REM (R)

Until 1953, it was assumed there were only these first four stages of sleep, with stage 1 accounting for about 33% of the total sleep period. Working in a University of Chicago sleep lab at that time, researchers Eugene Aserinsky and Nathaniel Kleitman were aware of early studies (~1922) reporting slow, rolling eye movements. Therefore, they placed electrodes above and to the side of each eye's orbit in 20 adult subjects. They observed four distinct periods of this eye-rolling throughout the sleep period, and they awakened the sleepers during these periods. They found that in about 75% of the awakenings, the sleepers reported "detailed dreams usually involving visual imagery" (Aserinsky & Kleitman, 1953, p. 273). Subsequently, Aserinsky and Kleitman were credited with the discovery of REM sleep. Further, although the cortical activity during the REM periods was similar to stage 1 sleep (low-amplitude, high-frequency waves), such that it was first called stage 1 REM, researchers came to realize it was a distinctly different stage of sleep.

Physiologically, REM sleep does not appear to be restful; heart rate and blood pressure fluctuate greatly. In men, REM is accompanied by penile erections (regardless of dream content), and women have a similar clitoral response. Thoughts, fleeting images, and ideas may occur in most sleep stages (though rarely in SWS), but highly detailed visual images, stories, and themes are almost exclusively associated with REM sleep. In human infants, REM sleep may account for up to about 75% of a total sleep period, but by adolescence the percentage drops to about 20 to 25%. This percentage remains fairly stable throughout the remaining lifespan, it declines only slightly after age 60 years. Phylogenetically, REM sleep is most prevalent in mammals and birds, although it has been observed in some lizards. Thus, REM's origins may be traced back at least 200 million years if not earlier, perhaps to the ancestors of some extant reptile.

One interesting characteristic of REM sleep is its accompanying muscle atonia (loss of muscle tone, or muscle "paralysis"). French sleep researcher Michel Jouvet (1972, 1980) and his colleagues (Rampin, Cespuglio, Chastrette, & Jouvet, 1991) explored the role of inhibitory neurons on voluntary muscle systems in preventing dreams being acted out by sleepers. They also proposed that REM sleep helped to preserve the integrity of one's psyche, based in part on studies showing the devastating psychological effects of REM deprivation in humans and other animals. Morrison (1983) demonstrated that selective destruction of these inhibitory neurons allowed cats to act out predatory actions, which presumably represented the content of their REM dreams. There are numerous anecdotal reports of people becoming aware of this muscle paralysis during REM sleep, and their accompanying dream themes often reflect the interpretation of muscle atonia, such as being paralyzed by aliens or being crushed by ghosts. Wing, Lee, and Chen (1994) found that over 93% of 603 Chinese undergraduate students had heard of the ghost crushing dream and 37% claimed to have experienced it. It has also been hypothesized that one of the most common themes of adult dreams, falling, may in part occur because of the sleeper's interpretation of the complete muscle atonia that accompanies the onset of REM sleep (e.g., Van de Castle, 1994). The other suspicion, of course, is that the falling dream theme is connected to our arboreal primate origins, as falling out of a tree was an event that an early primate would not have taken lightly nor easily forgotten (e.g., Sagan, 1977). Some modern sleep researchers have even called it a third state of consciousness, postulating wakefulness, sleep, and REM sleep. This trinity is actually an ancient one: early Hindu writings depict vaiswanara (wakefulness), prajna (dreamless sleep), and taijasa (dream sleep).

It has also been suggested that reports of alien abductions and alien experiments on the abducted are REM dreams accompanied by muscle atonia (Coolidge, 2006). Studies of people who report alien abductions are often no different on any standard measures of personality than those who do not report them. They appear only to differ in their strong belief in the existence of aliens. In my opinion, Freud was right: some dreams are wish-fulling (Freud, 1900/1956 [he claimed all dreams were wish-fulfilling but later modified that stance]). Interestingly, these alien abduction reports are invariably said to have occurred during the night (when people are most likely sleeping and dreaming). Thus, it may be that people who report alien abductions are having their wishes fulfilled by such dreams, and their paralysis during the alien experiments may be explained by the muscle atonia during REM. I shall not speculate further on reports where aliens probe the dreamer's genitalia or other orifices.

Major Theories of Why Organisms Sleep

Reduced Energy Demands

There are a number of theories for why sleep evolved, and I have hinted at a number of them already. It is important at the outset to reiterate that sleep and its stages have evolved for more than one reason. It can be said that sleep's evolution has been overdetermined (i.e., sleep is caused by many factors). As noted previously, there may have been many prominent evolutionary adaptations affecting sleep and many subsequent exaptations of sleep and its stages. Thus, it may be a little misleading to ask, "What is the purpose of sleep?" because the question implies there is only one reason. First, there are probably multiple reasons for the evolution of sleep and its stages, resulting in a multitude of proposed sleep phenotypes (i.e., physical and behavioral). Second, even if there was a single evolutionary purpose for sleep, it may have changed over the 600 million years of animal evolution.

Nonetheless, one prominent theory has always been that sleep reduces energy demands. An organism that operates continually, without rest, has greater energy requirements than one that has periods of rest. There are species which do not rest, like most species of sharks, but their continual activity comes at a cost. Their cortices, which normally would have greater metabolic needs than other types of tissue, are extremely small for the volume of their bodies. This makes sharks highly reflexive animals, and sharks whose cortices have been carefully ablated (surgically removed) can still swim, breathe, and feed for many days before dying. Some air-breathing sea mammals (cetaceans, such as porpoises and whales) have much larger brains (for their body size) than sharks' brains, making them much less reflexive. Further, cetaceans have managed to evolve a system that allows their expensive metabolic brain tissue to alternately sleep and be awake. As noted earlier, in porpoises one hemisphere will sleep while the other is awake, and this occurs about every 15 minutes. Interestingly, too, since this alternative-hemisphere sleep will result in visual neglect of one visual field, porpoises usually move side to side in their pod for blind-side protection.

The two animal lineages with which we share a common ancestral brain, mammals and primates, exhibit SWS and REM sleep. Moreover, those stages often occur in proportions similar to that in modern humans. The only major difference between their sleep and human sleep is whether they have nocturnal or diurnal sleeping patterns. SWS and the lighter stages of sleep appear to involve fewer metabolic demands, so their evolution is consistent with the reduced energy demands hypothesis. REM sleep in most species requires a greater expenditure of energy than do the other stages. However, as will be discussed shortly, there may be other evolutionary advantages of REM sleep that outweigh its increased energy costs and increased vulnerability to predation that accompanies REM sleep muscle atonia.

The Facilitation of Learning and Memory by Sleep

In a classic sleep study, two Cornell professors (Jenkins & Dallenbach, 1924) demonstrated that memory for a list of words was better after a period of sleep than after a period of wakefulness. They concluded, "Little is forgotten during sleep, and, on waking, the learner may take up the task refreshed and with renewed vigor" (p. 611). From their findings they interpreted forgetting as being a matter of "interference, inhibition, or obliteration" (p. 612) during waking and the protection of learned material during sleep. Subsequent empirical studies supported their findings; however, many of these studies used a sleep deprivation paradigm, which is problematic because the observed memory disruption may have been solely due to sleep deprivation and not to some more intimate connection between the stages of sleep, the sequence of stages, or the integrity of a single sleep period. Studies over the past two decades have demonstrated clearly that the relationship is more profound and that sleep indeed not only consolidates memories but also enhances them (e.g., Diekelmann & Born, 2010; Siegel, 2009; Tucker, Nguyen, & Stickgold, 2016). Sleep may also prune or eliminate irrelevant recent memories from further storage, allowing only more relevant and pertinent memories access to long-term storage (Stickgold & Walker, 2013).

Early research in this area revealed that REM enhanced only procedural memories, such as maze running in rats (e.g., Winson, 1990). Walker (2005) provided evidence that procedural memory consolidation and enhancement depended on SWS and other stages as well as the pattern of sleep stage sequencing. Indeed, there is evidence for procedural memory enhancement not only for maze running but also for visual and motor skill learning (e.g., Brawn, Fenn, Nusbaum, & Margoliash, 2010). There is also recent evidence that sleep may play a role in enhancing motor skills associated with music (i.e., repetition of learned finger sequences; Tucker et al., 2016). A variety of studies have demonstrated that declarative memories (dependent on the hippocampus for encoding) were also consolidated and enhanced during sleep (e.g., Diekelmann & Born, 2010; Marshall & Born, 2007).

Stickgold and Walker (2013) proposed that a sleep-dependent memory triage exists such that memories created during waking are subsequently chosen for offline (sleep) processing based on their waking "salience tags." That notion is completely consistent with my contention in Chapter 2, that memories (particularly of the episodic kind for what, where, and when) are a strong function of their emotional valence. Specifically, if a word, thought, idea, or experience is emotionally arousing, based on its inherent nature, one's motivation, or one's arousal level, then that memory will persist. If the memory lacks those properties, it will most likely be forgotten. According to Stickgold and Walker, memories that are selected and reactivated during sleep join other recently formed memories and are integrated such that "*de novo* knowledge [is created] beyond that available from individual item memories" (p. 144). If they do not join other memories, the "sleep-dependent processing may be limited to the comparatively straightforward consolidation and enhancement of the recently encoded item-memory itself" (p. 144). Evidence for this de novo knowledge, and even creativity, comes from studies by Cai, Mednick, Harrison, Kanady, and Mednick (2009); Wagner, Gais, Haider, Verleger, and Born (2004); and others.

Sleep and the Synaptic Homeostasis Hypothesis

A third reason for sleep comes from a proposal by Tononi and Cirelli (2014) that sleep is the price that a brain pays for neural plasticity. They hypothesize that synapses are strengthened while awake by learning and the formation of memories. However, this strengthening of synapses places extra demands on the neurons involved and their supporting glia. As a result, brain cells are under "cellular stress" (p. 12) during learning and memory periods. During sleep and its disconnection from the external environment, synaptic strengths are thought to renormalize, which restores their subsequent ability to learn and enhance signals in signal-to-noise ratios. In support of their theory, others have demonstrated that a multitude of genes change expression during sleep and during particular sleep stages (such as REM) in humans and a variety of the other species (e.g., Mikhail, Vaucher, Jimenez, & Tafti, 2017; Ribeiro, Goyal, Mello, & Pavlides, 1999; Zhang, Lahens, Ballance, Hughes, & Hogenesch, 2014). Further support for their theory that sleep is the price paid for neural plasticity comes from the finding that sleep and the percentage of REM sleep is greatest during the period of life when neural plasticity is at its greatest, that is, infancy and early childhood.

Related to neuronal explanations for sleep, it is clear that DNA provides instructions for sleep and wakefulness. However, sleep's proximate physiological basis appears to be the removal of metabolites (a cell's waste products) that accumulate during wakefulness (Xie et al., 2013). Xie and his colleagues have argued that it is sleep that facilitates this clearance. However, the nature of the metabolites varies, as some are toxic and some, like adenosine, induce sleep. Herculano-Houzel (2015) hypothesizes that sleep-inducing metabolites depend on neuronal density and a brain's total surface area. She also claims that such brain changes (i.e., increasing neuronal density and increasing surface area) helped decrease the percentage of sleep and increased feeding activities in both mammalian and *Homo sapiens*' evolution. Again, however, it is important to note that sleep's architecture is over-determined, and ratios of sleep to wakefulness are much more likely to be a function of multiple factors.

Priming

Traditional cognitive priming effects are said to occur when prior exposure to stimuli enhances or changes later recognition and performance, regardless of whether there was conscious recognition of the stimuli. These traditional priming effects have been demonstrated in sleep and learning studies (e.g., Stickgold, Scott, Rittenhouse, & Hobson, 1999). However, I am referring to priming effects from the contents of thoughts and dreams that occur during sleep and REM, which as discussed earlier can be de novo creations. This is my fourth reason for the evolution of sleep and dreams: Priming can occur even from "imaginary" stimuli like dreams and sleep thoughts. Finish psychologist Antti Revonsuo (2000) proposed a *threat simulation hypothesis*, which states that the major biological function of dreaming is to simulate threatening events. He found support for his hypothesis from the preponderance of children's scary dreams, nightmares in children and adults, dreams after trauma, and the heightened aggression dreams of modern hunter-gatherers. He thinks the environment of ancestral humans was "short and full of threats" (p. 793); therefore, any advantage that might be gained from rehearsing these threats in dreams before waking would likely have put those humans at a relative fitness advantage (e.g., survival and reproduction).

Franklin and Zyphur (2005) extended Revonsuo's hypothesis to propose that dreams may be a reflection of a more general virtual rehearsal mechanism, which may enhance the success of subsequent waking endeavors and even influence subsequent behavioral predilections. As Franklin and Zyphur pushed the boundaries of their virtual rehearsal mechanism, they noted that any simulations during dreaming had to be perceived as real, otherwise they might be dismissed as simply "dreams," which would not carry the subsequent emotional valence necessary for success in waking life. Indeed, as noted in Chapter 3, the dorsolateral prefrontal cortex is relatively inactive during REM, which might otherwise allow dreams to be dismissed as unreal and unimportant. They also interpreted the seeming randomness of dreams and their sometimes bizarre content and often surprising scenarios as potentially adaptive, because dreams may provide "a broad range of scenarios to be simulated and new scenarios to be created rather than having the same type of dream occur repeatedly" (p. 67). This idea is entirely consonant with Hartmann's (1998) idea that dreams make broader connections between ideas than do waking minds, and often dreams are guided by emotional valences and concerns. He states that "dreaming avoids the 'tightly woven' or 'overlearned' regions of the mind (such as those concerned with reading, writing, and arithmetic)" (p. 3). Further, he proposed that new material may be processed in dreams such that it becomes interwoven within existing memory connections, and the latter, he proposed, is highly useful in solving novel problems, in scientific endeavors, and in artistic endeavors (which I shall review shortly).

Revonsuo, Tuominen, and Valli (2015) later expanded these ideas beyond the threat simulation hypothesis and proposed a *social simulation theory* of dreaming, which posits that dreaming simulates social interactions, skills, and bonds that might be encountered during waking. They referred to the human elements in dreams as *avatars* and hypothesized that the avatars force dreaming humans to "maintain and practice various evolutionarily important functions of social perception and social bonding" (p. 28).

Creativity

A fifth reason for the evolution of sleep and dreams also derives from the contents of the thoughts, ideas, and vivid imagery that arise when humans sleep. The first preserved writings of historical civilizations like those of the Mesopotamians and Egyptians are replete with references to dreams (see Van de Castle, 1994, or Coolidge, 2006, for reviews of this literature). More recently, there have been numerous anecdotal reports in the arts about inspirations coming from dreams. Artists who have claimed that ideas for a painting came from a dream include Salvador Dali, Francisco Goya, René Magritte, and many others. One of Picasso's most famous pictures, *Le Rêve* (*The Dream*; it is a surrealistic, dream-like picture of his mistress) may or may not have been inspired by a dream, although its title certainly was. South African archaeologist David Lewis-Williams (2002) has proposed that Upper Paleolithic cave art (~30,000 years ago) must have been inspired by shamanistic trances and hallucinatory states induced by drugs from plants. In my opinion, however, he dismisses or overlooks a much more common source for imaginative cave imagery: dreams, and in particular, therianthropes (half-human, half-beast). More importantly, dreams would have been a more readily available and steady source of inspiration to everyone, not just those who used drugs.

Musicians who claim some of their inspirations came from dreams include Wolfgang Mozart, the German composer Richard Wagner, Billy Joel, and many others. One particularly compelling story of a musical piece inspired by a dream comes from the life of Giuseppe Tartini, an Italian composer and violinist, who said the devil appeared to him in a dream. In this dream, Tartini handed the devil his violin, and the devil played a song with such expertise and beauty that upon awakening, Tartini composed his now famous Violin Sonata in G minor or "Devil's Trill Sonata." This sonata and story undoubtedly inspired country-rock musician Charlie Daniels to compose the popular 1979 song "The Devil Went Down to Georgia."

Writers have long claimed inspirations come from dreams, and many of these works have been highly influential for almost two centuries. For example, Samuel Taylor Coleridge wrote his classic fragmented poem *Kubla Khan* in 1897 (not published until 1916). He reported his inspiration for it came from a dream, although to what extent it was also opium influenced is unknown. Mary Shelley published *Frankenstein* in 1818, and she claimed that it was based on her dreams, one of which included the revival of her child that had died shortly after childbirth, and another of a scientist who learned to create life and was horrified by his creation. In 1866, Robert Louis Stevenson published *The Strange Case of Dr. Jekyll and Mr. Hyde*, which was a battle of good and evil within a person suffering from what is now recognized as dissociative identity disorder (i.e., multiple personalities). Stevenson claimed that his inspiration came from one of his nightmares. Modern writers such as Steven King also claim that many of their ideas come from dreams. In particular, King said the idea for his 1987 book *Misery* (later a movie) came from a dream while napping on a transatlantic flight.

Scientists also claim that many of their inspirations come from dreams. One of the most famous dream stories linked to a scientific discovery is attributed to German chemist August Kekulé (1829–1896), who, in 1865, was the first to publish on the nature of the ring shape of the benzene molecule. Many years later, he claimed that the idea for the ring shape came during a reverie or daydream of a snake seizing its own tail. As Russian chemist Dmitri Mendeleev (1834–1907) was working on a system to classify elements based on their chemical properties, he reported that the periodic table came to him in a dream. He wrote that, upon awakening, he wrote his vision down on a piece of paper, and he said it subsequently required only one correction. The brilliant Indian mathematician Srinivasa Ramanujan (1887–1920) claimed that his theorems and equations were given to him in his dreams, from his family's favorite goddess Mahalakshmi. Anecdotal reports and interviews of contemporary mathematicians reveal that a majority say that they solved a problem in a dream or thought it was a likely possibility (e.g., Coolidge, 2006; Krippner & Hughes, 1970). When I sought further evidence from a mathematics professor at my university, he said that had never solved a math problem in a dream. He then paused, and said, "But I dream of the most wonderful problems!"

In this review of wondrous discoveries whose origins may have stemmed from dreams, it is important to note that virtually all reports come from people who were experts in their field or craft. Moreover, their dreams' ideas came only after intense work and long hours of expert labor while awake. Finally, I suggest that it may be important to this "inspiration from dreams" hypothesis that waking and sleep be viewed as a continuum of consciousness, where waking experiences, thoughts, and ideas are often carried over into sleep and vice versa. Thus, after years of intense, conscious work done while awake, it might be deemed highly unlikely that these ideas would not be extended into sleep's peculiar levels of consciousness. Recently, my colleague archaeologist Thomas Wynn and I (Coolidge & Wynn, 2018) pondered whether the design for the Acheulean handaxe, a symmetrical, leaf-shaped stone tool that first appeared about 1.9 million years ago, came in a dream to an expert stone tool knapper. The design was apparently so useful that it persisted for at least the next 1.5 million years or more. Perhaps the original design came from a dream of a falling leaf, whose shape was then imposed on a stone.

An Empirical Test of Sleep Insight and Creativity
German psychologist Ullrich Wagner and his colleagues (2004) published one of the first empirical tests of the potential for creativity during a sleep period. In their study, 106 adults were given a cognitive task that required

learning stimulus–response sequences (an implicit procedural memory task). Improvement was measured by decreasing reaction times. Participants could improve abruptly if they gained insight into a hidden abstract. Twice as many participants who slept for 8 hours after training became aware of the hidden rule than those who stayed awake for 8 hours after training. The authors hypothesized that the participants' greater insight was not a strengthening of the procedural memory itself but an actual novel restructuring of the original representations. They speculated that the restructuring was mediated by the hippocampus, related medial temporal lobe structures, and the prefrontal cortex. These structures have been shown to play an important role in generating awareness in memory. Wagner et al. suspect that cell assemblies encoding newly learned tasks were reactivated during sleep by hippocampal structures, which were then incorporated into pre-existing long-term memories by the neocortex. They hypothesize that this process of incorporation into long-term storage formed the basis for the remodeling and that it was a qualitatively different restructuring of representations in memory. Thus, in their opinion, sleep may serve as a catalyst for insight.

Sleep and *Homo erectus* Revisited

As noted in Chapter 1, a major grade shift (i.e., when a group develops completely new behaviors and adaptations) occurred with the appearance of *Homo erectus* about 1.9 million years ago. There are at least three major reasons for this relatively sudden change in the evolutionary trajectory of the *Homo* species. The genus *Homo* began with *Homo habilis* about 2.5 million years ago, which had a clear association with the first stone tools (i.e., cores, hammer stones, and sharp flakes) and a near doubling in brain size. Despite stone tool manufacture and much larger brains, the habilines appeared to maintain their largely arboreal lifestyle and restricted range, just like the australopithecines. The latter supposition is largely based on habilines' limbs and body proportions, which were australopithecine-like (longer limbs, shorter and lighter bodies) and thus designed for a life in trees. Moreover, archaeologists had originally thought the first stone tools were only associated with *Homo habilis*, but there is recent evidence that australopithecines might have occasionally made and used sharp flakes. In 2015, 3.34 million-year-old stone tools were discovered in Lomekwi, West Turkana, Kenya, and it appeared that the knappers had some understanding of how stones fractured and how to make sharp flakes from a core. Since *Homo habilis* did not appear until about 800,000 years later, this meant that some hominid much earlier had the same stone tool–knapping abilities of *Homo habilis*. However, to date, no clear hominid skeletons have been found associated with these stone tools, but it is suspected that they may be attributed to the australopithecines. The implications of this find are interesting. Since the genus *Homo* was ascribed to the species *habilis* because they were thought to be the first to make stone tools, either the genus *Australopithecus* should now be changed to *Homo* or the hominin *Homo habilis* should be changed to *Australopithecus habilis*. Another remote possibility is that the Lomekwian stone tools are the product of *Homo habilis*, but there is currently no evidence that the habilines emerged that early.

There is no question, however, that *Homo erectus* was a major leap in the evolution of hominins. Beginning a little more recently than 2 million years ago, *Homo erectus* was much larger physically than the earlier australopithecines and habilines. *Homo erectus* was almost 2 meters tall (about 6 ft 1 in.) and weighed about 54 to 64 kilograms (120 to 140 lb). This means that *erectus* had made the complete transition to terrestrial life and no longer lived in trees. Their brains were also larger (about 950 cc [range 850 cc to 1,100 cc]) relative to their larger and heavier bodies, with an EQ of about 5.0, whereas the australopithecines' EQ was about 2.2 to 2.5 (modern humans' EQ is about 7.5).

The second reason for the sudden change in the evolutionary trajectory of *Homo erectus* is that their stone tools showed a dramatic increase in symmetry, beauty, and complexity. They were the first to produce a stone handaxe, which, as noted earlier, was leaf-shaped and about the size of one's hand. It was clearly knapped from a larger flake, which means *Homo erectus* was able to manage the core properties of a larger stone, prepare it in order to knock off a large flake, and then shape the flake into a sinuously edged, symmetrical (biface) handaxe, which then could be used to smash, cut, scrape, and otherwise prepare various foods for consumption. This dramatic increase in symmetry, beauty, and complexity of stone tools was also undoubtedly accompanied by a greater load on visual working memory, but it did not necessarily require the characteristics of "enhanced" working memory, which we have only attributed to modern *Homo sapiens* (Coolidge & Wynn, 2018; Wynn & Coolidge, 2016).

Perhaps *erectus'* greatest challenge (and thus the environmental hurdle shaping the species' adaptations) came from living on the ground. Because the australopithecines and habilines lived and slept in trees for protection from predation (although they did spend time on the ground foraging and scavenging for meat), there had to have been a limit to how many individuals could nest in a tree, otherwise an excessive number of nests would probably attract predators. Extant chimpanzees, whose average group size is about 40 members, also make nests and sleep in trees. Interestingly, when a chimps' group size becomes upwards of 80 or so members, the group will split into two. As life on the ground obviously entailed a greater risk of predation, *Homo erectus* groups must have been larger than those of the australopithecines or habilines. Based on the work of anthropologist Robin Dunbar (1998), it may be surmised that *erectus* group sizes were much larger than those of australopithecines and habilines, and anatomically modern humans may have consisted of groups of about 150 members.

Yet, imagine the other challenges besides the risk of predation from living and sleeping on the ground. Nests in trees had to be replaced by some kind of shelter on the ground. Although most anthropologists eschew the general notion of "cavemen" (and cavewomen), it is highly likely that rock overhangs and small caves were often sources of shelter for *Homo erectus*, and a location near a water source like a stream or river was likely. Given *Homo erectus'* stature, a body seemingly designed to run, and probably run long distances (Bramble & Lieberman, 2004), a convenient water source was an absolute necessity. The rock and stone materials they used for knapping often came from long distances, so anthropologists estimate that their territorial expansion (about 260 square kilometers. or 100 square miles) was probably about 10 times the expanse of the australopithecines or habilines. Perhaps even more provocative is that within just a few hundred thousand years of their first appearance, *Homo erectus* groups began to expand to other parts of Africa, the Middle East, Europe, Southern Asia, India, and Southeast Asia and Indonesia. As noted earlier, Cachel and Harris (1995) referred to *erectus* as a "weed species" because *erectus* spread far and wide but never in heavily dense numbers. They also suggest that *Homo erectus* was especially adept at invading "disrupted" environments, beginning about 1.7 million years ago. This characterization is supported by Wells and Stock (2007); however, they focused on *erectus'* increased body size and culture and dismiss any causal relationship from an increase in brain size. As I shall discuss shortly, biology has strong links to culture, and dismissing the influences of brain size and brain shape changes on culture is myopic.

In my opinion, there is a common flaw in many cultural (learning-based) explanations for phenotypes. As American biologist E. O. Wilson (2012) noted, biology and genes place a leash on culture. Certainly, culture has a reciprocal influence on biology and genes, but to *dismiss* the increase in brain size in *erectus* (and increased encephalization quotient) as an influence on the grade shift in behavior of *erectus* is particularly disingenuous. Perhaps the most vivid counterexample is that of chimpanzees raised from birth in a modern human culture. No matter how many humans they interact with or how many years they are exposed to modern human culture, they cannot read, write, or speak, although they do "respond" to human language. Observations of their learning some arbitrary signs representing tangible things, like foodstuffs and other animals, and responding to some of modern human language's sounds can simply be attributed to classical and operant conditioning, which, as noted earlier, can be observed in flatworms with rudimentary brains. The supposition that chimpanzees can "understand" human language is simply unjustified, but that they respond to human language is not. Related to this issue, see my recounts of Nagel and Wittgenstein's comments at the end of Chapter 10. In summary, chimpanzees' primary ways of behaving remain firmly "leashed" by their genes and their inherent biology. Thus, for Cachel, Harris, Wells, and Stock to view *Homo erectus'* dramatic grade shift in behavior compared to the australopithecines and habilines as being simply due to an increase in number of group members is to put the cart before the horse. No amount of human rearing and exposure to human culture has allowed chimpanzees to speak and understand language even at the level of a human 2-year-old. Biology has placed a very tight leash on their culture.

The transition from tree to ground life challenged hominins in a myriad of ways. Providing shelter and increasing group membership as a defense against predation made highly novel demands. Variations in learning abilities and being behaviorally more flexible helped select for the faster, smarter, and more adaptable hominin, which is a classic Baldwin effect, as noted in Chapter 2. Moreover, the fitness advantages accrued by these adaptations indicate that the genes accounting for them were passed on to their offspring. As bigger brains required more calories, a switch to a meatier diet may have also played a role in their rapid expansion. As noted previously (DeLouize, Coolidge, & Wynn, 2016), meat increases levels of the neurotransmitter dopamine, which plays a profound role in motor behaviors and in some important cognitive-behavioral innate predilections like exploratory behavior. It is

THE EVOLUTION OF SLEEP AND DREAMS 199

not a mere coincidence that a host of novel behaviors appeared with the evolution of *erectus*: the management (if not production) of fire, the all-purpose
stone handaxe, long-distance running, and most likely a change from the
mere scavenging of meat to hunting meat. Thus, this transition to terrestrial
life was most likely also a change from being prey to becoming a predator.

Cognitive Repercussions

So far, I have resisted interjecting comments about the cognitive abilities of *Homo erectus*. From my review of learning and memory systems in
Chapter 2, *erectus* must have had both non-associative (habituation and
sensitization) and associative (classical and operant conditioning) learning
abilities. Further, because Allen and Fortin (2013) grant episodic memory to
animals even hundreds of millions of years ago, then *erectus* must have had
episodic memory capabilities. The only question I would have of this ability
is to what extent their episodic memory was autobiographically influenced.
Did *erectus*, particularly about 1.7 million years ago, with the advent of the
true Mode 2 handaxe and the ability to populate Southwest and Southeast
Asia and India (by 1.5 million years ago at Attirkappam, India), have a sense
of self in their episodic memories? Further, as I have noted previously, if
modern vervet monkeys have different sounds for predators and modern
putty-nosed monkeys can combine sounds that have a new meaning, then
surely *Homo erectus* vocally communicated with one another, and these
communicative abilities were probably unmatched by any previous hominid. There are numerous studies that support the contention that *erectus* did
have the vocal and breathing apparati to produce language-like sounds (e.g.,
Walker & Leakey, 1993). But did *erectus* have real language? Whatever the
limits of their cognitive abilities, I think I have made the case that *erectus*
truly made the transition from an ape-grade way of life to a more human-
grade way of life.

Discussions of language and even its basic definition, as noted in
Chapter 6, are, however, highly contentious. However, a majority of linguists
do believe that there were protolanguages before modern human language
(e.g., Bickerton, 2007; Botha, 2002, 2006, 2012), and many accord *erectus* a
protolanguage. Remember also that *erectus'* brain was about 65% the size
of modern human brains when the species first appeared about 1.9 million
years ago. Later, about 1.5 million years ago, and after the appearance of

erectus' complex handaxe, their brains approached 80% of modern human brain size. Of course, as I have made clear throughout, complex cognitive functions arise not only as a function of brain size (even relative to increases in body size) but also as a function of its shape—the shape of *erectus'* brain was even more consequential than simple EQ. Paleoneurologists have shown that the shape of *erectus'* brain was changing (e.g., Bruner, 2004, 2010; Bruner & Holloway, 2010) compared to that of the australopithecines and (likely) the habilines, again implying a change from being ape-like to more human-like. Bruner and Holloway documented that *erectus* had increased lateral portions of both the temporal lobes and the inferior portions of their parietal lobes compared to the australopithecines. They also suspect that there may have been slightly wider frontal lobes in *erectus*. As previously noted in Chapter 3, all of these brain regions have played prominent roles in modern human cognition. The frontal areas are involved in decision-making, sequencing, elements of theory of mind, and the guidance of stone knapping. The temporal lobes have a major role in language comprehension (or the application of meanings to sounds and coupled with the frontal lobes' speech production). Finally, the inferior parietal lobes, which house the supramarginal gyrus and the angular gyrus, have critical roles in inner speech and the multimodal, higher-level processing of cognition and aspects of mathematics. The idea that the anterior portion of the supramarginal gyrus may have begun to be under positive selection in *erectus* as well is highly intriguing. Orban and Caruana (2014) have shown through brain imaging that the neural activation patterns of modern monkeys (macaques) and modern humans is identical when the former are manipulating objects and the latter are using tools. One of the few differences in activation was in humans' anterior portion of the supramarginal gyrus, which is involved in the sequencing of actions, and it receives input from the fusiform gyrus about an object's shape. Although Bruner and Holloway's work cannot confirm the specific supposition that the anterior portion of the supramarginal gyrus was enlarged in *erectus* compared to that in the australopithecines, it is interesting that it may have begun expanding, which certainly would have aided the creation of the more complex Mode 2 handaxe (which will be discussed shortly). It is important to note, however, that there are still substantial differences between *erectus* brains and later *Homo* brains, and these differences probably heralded some significant cognitive differences as well.

Another contentious language division is whether modern language appeared suddenly and whole cloth (saltationism) as Chomsky and his colleagues purport (Bolhuis, Tattersal, Chomsky, & Berwick, 2014), or did it evolve gradually from protolanguages? As I hypothesized in Chapter 6, both sides are partially right and partially wrong. Certainly, Chomsky's stance that there was no language prior to 100,000 years ago is absurd, and there is no genetic evidence whatsoever that language occurred in one person because of a single genetic mutation. Unfortunately, because of Chomsky's prior reputation in linguistics, his poorly reasoned contentions have inordinate influence. Nevertheless, it does appear that something happened around 100,000 years ago which makes most scientists think that the cultural explosion beginning around 50,000 years ago (cave art, figurines, highly ritualized burials, personal ornamentation, etc.) might have been a product of a more sophisticated language system. My colleague archaeologist Thomas Wynn and I attribute this cultural explosion not to language per se, but to a genetic or epigenetic event that enhanced working memory capacity (Coolidge & Wynn, 2001, 2016, 2018; Wynn & Coolidge, 2010, 2016). Repeatedly, we have been misquoted and misconstrued in the literature as having said a "single" gene gave *Homo sapiens* modern minds. First, a single gene *can* have an inordinate influence on the brain and, presumably, cognition. Evans et al. (2005) found a single gene locus, *Microcephalin* (MCPH1), that appears to have enhanced brain size about 37,000 years ago and spread quickly in the human population under strong positive selection, and it continued to evolve rapidly. Recently, a search for other genes that may have driven larger brains much earlier revealed that a group of genes, called *NOTCH*, were identified in human fetal tissue. This small group of genes appears to be directly involved in the rapid growth of human fetal brains. Subsequently, these researchers found that specific genes were involved in sending information between cells and that they were present in apes, humans, and our extinct cousins, the Neandertals and Denisovans. Importantly, they were not found in chimpanzees, so the gene-copying DNA process that created these genes might have occurred sometime after the divergence of chimpanzees and the lineage of modern humans about 6 to 10 million years ago.

Thus, arguments about the plausibility of a single-gene event in the creation of modern brains do bear some consideration, although it is not the genes per se that have positive selection value but the phenotypes (i.e., behaviors) they control. Wynn and I never proposed a single-gene event hypothesis. What we did propose was that some genetic event did occur that enhanced some cognitive ability or function (i.e., executive functions, working memory capacity, phonological storage, etc.) sometime more recently than about 200,000 years ago. But I digress. The latter discussion focuses on an issue long after *Homo erectus* evolved.

I propose that *Homo erectus* had some form of protolanguage with many different sounds for many different objects and ideas in their world. Further, I have the strong intuition that they shared many of the overt pragmatics (reasons for speech) that modern humans possess: I think they could warn others and express their emotions (exclamatives), like alarm calls and screaming in pain and pleasure. I think they could use sounds to exert their dominance and will over submissive others (imperatives), like telling younger members to move or yield. I think they could name things (declaratives), like having consistent sound-names for the sun, water, or meat. Less possible but still possible, I think they could pose questions (interrogatives), like asking where someone eating meat got the meat or where someone obtained some colored stone. As I noted previously, all four of these pragmatics can be expressed in English and many other languages in a single, simple syllable: "Ouch!" "Move!" "Sun." "Where?" Remember, too, that Dunbar (1998, 2013; Dunbar & Schultz, 2007) proposed that social grooming replaced physical grooming, which also became a sine qua non in the expanding group size of *Homo erectus* and their terrestrial life. Social grooming and physical grooming have been shown to reduce aggression and enhance social bonding.

The only pragmatic of speech of which I am very doubtful that *erectus* possessed is the subjunctive mode of speech, that is, "what if" statements and ideas contrary to fact. And to return to my digression just briefly, I propose that the subjunctive pragmatic of speech was released when some genetic or epigenetic event occurred in the past 200,000 years that enhanced executive functions or working memory capacity in some fashion. I think subjunctives require a suite of higher cognitive abilities, such as increased phonological storage (to hold "whats" long enough in mind to apply "ifs" or to hold "ifs" long enough in mind to apply "thens") and higher levels of theory of mind ("If I say this to her, will she react positively, or should I say it this way?"). I think subjunctive thinking is or requires recursive thought. And here's where Chomsky and the language saltationists may be partly right: a genetic event may have occurred around 200,000 years ago or more recently that enhanced working memory capacity in some fashion that in turn made language fully modern. If the hallmark of modern language is recursion (embedding a phrase within a phrase; see Coolidge, Overmann, & Wynn, 2010 for an overview of recursion) and recursive thinking is a sine qua non for subjunctive speech, then indeed, the archaeological consequences of recursive modern language might be reflected in cave paintings, figurines, and highly ritualized burials.

Archaeological Evidence for More Complex Thinking in *Homo erectus*

It is perhaps unwise to dismiss the cognitive implications of the Mode 2 handaxe that first appeared with *Homo erectus* about 1.7 million years ago. I earlier noted the possible expansion of the supramarginal gyrus in *Homo erectus* (Bruner & Holloway, 2010), which ample evidence demonstrates to be activated (especially its anterior portion) when modern humans are stone knapping. Given that the anterior supramarginal gyrus is involved with sequencing and receives input from the fusiform gyrus for shape recognition, it is entirely plausible that this network (frontal, parietal, and temporal lobes) was responsible for the sophistication of the Mode 2 handaxe.

Recently, my colleague Thomas Wynn and archaeologist John Gowlett (2018) outlined the essential features of the handaxe (i.e., Acheulean handaxe, after Saint-Acheul, France, where it was first recognized; see Fig. 1.3). First, they claimed that it was ergonomically shaped—it was an efficient tool and fit comfortably in one's hand. Interestingly, its shape has been labeled a *hemilemniscate*, which means it is one-half of the symbol for infinity (a common leaf shape). The latter implies that the Mode 2 handaxe had its base of gravity in its bottom and that it thinned toward the sides and its tip. Second, this handaxe extended forward, unlike simple Mode 1 stone choppers (which were rounder and smaller), which provided a much longer cutting surface on both sides. From the side, it is easy to see the sinuous cutting edge, where small pieces were knapped away alternately from both sides as the knapper worked along the edge. One of my former students, Klint Janulis, an expert knapper and Paleolithic-skills instructor, made me a replica from UK flint, and its sinuous edge cuts easily through cardboard boxes, especially the heavy tape that binds them. Yes, it has been demonstrated that a dead elephant (unfortunately) can be stripped of its meat with just small sharp flakes, but this is highly inefficient and time-consuming compared to doing it with a handaxe. The alternately knapped sides also produce a lens-shaped curvature on both sides when the handaxe is viewed from its bottom.

Wynn and Gowlett (2018) proposed that the handaxe could not be too narrow as it extended toward its tip, as it would be unstable and might twist with hard use. They proposed that this feature, along with its weight, was also an ergonomic function and was created by the knapper by monitoring the handax's thickness. Finally, a most interestingly characteristic is that most handaxes show a slight skewness (either slightly tipping to the right or left). This characteristic reflects a handedness bias. It appears that even the earliest handaxes show that a majority of *erectus* stone knappers were right-handed, and more recent handaxes show a more pronounced bias for right-handed knappers. This implies that even the earliest form of *erectus* preferred their right hands. However, little or nothing can be made of this bias with regard to language lateralization as hand preference and language lateralization were probably independently derived traits (Fitch & Braccini, 2013). The strongest counter-evidence to claiming they are intertwined is that well over 95% of people who prefer their right hand have speech in their left hemisphere, but so do a strong majority of people who prefer their left hand.

Perhaps even more interesting is Wynn and Gowlett's (2018) proposal that handaxes were "over-determined," which in this archaeological context meant that their knappers spent more time than was needed to make a useful and ergonomically shaped handaxe. Not only did the knappers strive for sym-metry, and symmetry recognition is a cognitive ability in mammals, but what is unique is that these knappers imposed "symmetry on an artifact" (p. 9). An aspect of Wynn and Gowlett's "over-determinism" hypothesis is that the knappers may have enjoyed the finished product, which Wynn and Gowlett call the "aesthetic component." They note that even by 1.7 million years ago, not all, but some handaxes were overdetermined and aesthetically pleasing. Indeed, some handaxes were too big (gigantism) to be practical, and some appear to have been unused! The latter idea in particular has led to the highly provocative hypothesis that handaxes were a product of sexual selection; that is, a preponderance of stone knappers were men trying to impress women with their technical prowess to gain favorable attention as mates (e.g., Kohn & Mithen, 1999). In terms of the economics of sperm and eggs, the hypothesis makes sense. Sperm is cheap, as it is produced in the billions. Eggs, however, are precious, as a woman might produce only a few hundred in her reproduc-tive lifetime. Thus, sociobiologists have long theorized that this imbalance requires men to "ornament" themselves for attention, or in the present case, produce "ornaments" (i.e., handaxes) to be selected by women for reproduc-tion. This difference in reproductive strategies produces sexually dimorphic traits, perhaps not only physically but also behaviorally. Additional evidence that *Homo erectus* was acting in more human-like ways is the fact that the sexual dimorphism in terms of body size was reduced in *erectus* (as it is in modern humans) compared to the earlier australopithecines. The reduction in sexual dimorphism in *erectus* also suggests there was reduced aggression among men for women's attention.

Wynn and Coolidge (2016) declared that a "cognitive Rubicon" was crossed at about 1.75 million years ago with these Early Acheulean handaxes, where higher-order motor skills and cognitive abilities were a prerequisite for their construction. They propose that these skills and abilities promoted these hominins from ape-like thinking to more human-like thinking. Coolidge and Wynn (2019) recently proposed that a second cognitive Rubicon may have been crossed at about 500,000 years ago with the development of Later Acheulean handaxes. The archaeological evidence for this second cognitive Rubicon tells a dramatically different story from that of Early Acheulean handaxes in terms of spatial cognition, indexical possibilities, aesthetics, and social cognition. Specifically, the second Rubicon is marked by the much more frequent appearance of S-twist handaxes, ovates (where the equatorial axis exceeds or equals the polar axis, producing a spherical shape), handaxes with natural holes within their perimeters, handaxes with intentionally included fossils, the use of variegated stone material, zoomorphic handaxes that look like horse's heads or other animals, and overly large handaxes (gigantism). Wynn (2002) has proposed that the sophistication of these stone tools and their multidimensional symmetries might be indicative of mental "categories" (but not mental templates) that still might reside in long-term declarative/semantic memory.

Preliminary Summary

As stressed throughout this chapter, there appear to have been multiple purposes for the evolution of sleep and dreams, and if there ever was a single adaptive reason, it appears to have been lost in time. Nonetheless, sleep and its stages have been exapted innumerable times in evolution. In this chapter I reviewed six major reasons for the evolution of sleep and its stages: reduced energy demands, the facilitation of learning and memory, synaptic homeostasis, metabolic waste clearance, priming, and creativity. In addition, I argued that *Homo erectus* may have been the first hominid to have derived cognitive benefits from qualitatively better sleep *and* a single, integrated sleeping period. Certainly, Lucy (*Australopithecus afarensis*) and the habilines slept, had SWS and REM, and unquestionably had vivid dreams. However, I propose that *Homo erectus* was the first hominid to have relatively uninterrupted sleep, which preserves the integrity and benefits of SWS and REM cycles, especially on learning and memory and the replay of episodic memories during these periods. Further, the content and ideas in *erectus'* dreams were probably also present in the dreams of the australopithecines and habilines, but I hypothesize that *erectus* was probably the first hominid to transmit these dreams and ideas to future generations via culture and a protolanguage. I further hypothesize that *erectus'* protolanguage had many of the pragmatics of modern speech, including exclamatives, imperatives, declaratives, and interrogatives. All the benefits of sleep would have been enhanced because *erectus* slept on the ground, in shelters, and, most importantly to the evolution of modern cognition, in the protection of much larger social groups. The larger group size also placed demands on *erectus'* cognitive and social functioning, which favored larger brains, a differently shaped brain, and novel brain networks.

Summary

1. Primitive sleep states probably developed from earth's rotational cycle, where activities were constrained by alternating periods of light and dark.
2. Reptiles, birds, mammals, and primates exhibit slow-wave sleep (SWS) and rapid eye movement (REM) sleep.
3. There may have been no single purpose of sleep. Among the multiple reasons for its evolution and different sleep stages are reduced energy demands, facilitation of learning and memory (i.e., strengthening synaptic transmission), metabolic waste clearance, priming for subsequently successful waking activities, and aiding creativity.
4. When *Homo erectus* appeared about 1.9 million years ago, it made a full transition to terrestrial life, including sleeping on the ground instead of in nests in trees. Ground sleep resulted in a single integrated sleep period, which had many positive cognitive repercussions.

5. The design for the Mode 2 (Acheulean) handaxe, attributed to *Homo erectus*, may have come to a stone knapper in a dream. Some handaxes, at the time of *erectus*, had an aesthetic quality and may have been used as a means of sexual selection.

6. Early Acheulean and Later Acheulean handaxes differed in quality and aesthetics such that the later handaxes may represent the crossing of a second cognitive Rubicon, where even greater motor abilities and cognitive skills may have been prerequisites for their manufacture.

10

Paleopsychopathology

Paleopsychopathology is the study of mental problems and mental diseases that may have increased relative fitness (adaptive) in the ancestral environment but do not enhance fitness in the present environment (maladaptive). One initial assumption for this supposition is that mental problems and diseases are genetically based, and this appears to be an extraordinarily safe assumption as almost all types of psychopathology have a highly heritable basis (i.e., under tight genetic control). In fact, it is commonly thought that there is *no* complex human behavior that does not have an underlying genetic predisposition, even political leanings (e.g., Hatemi & McDermott, 2012; Turkheimer, 2000). With few exceptions, complex human behaviors and psychopathologies are influenced by a combination of many genes (polygenic), while some like autism and schizophrenia may be caused by hundreds of genes working in combination (as will be discussed shortly).

Evolutionary psychologists often distinguish between proximal and ultimate explanations for behavior. A *proximal* explanation involves mechanisms that are an immediate cause for behavior, such as physiological, hormonal, structural, developmental, or experiential. An *ultimate* explanation refers to the evolutionary adaptive value of a behavior that enhanced the relative fitness of an organism in the ancestral environment. At the outset, it is also important to note that not *all* psychopathology may have had adaptive features. The mental deterioration and severe cognitive disruption that occur with some pathologies (like Huntington's chorea) may not have had any adaptive value in the ancestral environment, as that particular genetic disorder (caused by a single dominant gene) can arise well after what would have been the prime reproductive years in the ancestral environment. Further, there is evidence that the power of natural selection wanes across the lifespan. It is also important to note that many of the present explanations for ancestral adaptations of extant mental problems are invoked at the individual level, but as Nichols (2009) has insightfully noted, evolutionary advantages can also be explained at the kin level (genetic relatedness among individuals) or at a group level (behaviors of an individual that are beneficial to the group).

Evolutionary Neuropsychology. Frederick L. Coolidge, Oxford University Press (2020).
© Oxford University Press.
DOI: 10.1093/oso/9780190940942.001.0001

Another factor to consider is that an individual psychopathology may not have been advantageous in and of itself, but rather its polygenic basis or its proximity to important genes may have given rise to successful adaptive phenotypes. For example, in a 2014 genome study of 150,000 people (37,000 of whom were diagnosed with schizophrenia), 108 genes were associated with a schizophrenia diagnosis. The DNA regions that showed the strongest association with that diagnosis also coded for proteins of the major histocompatibility complex, which are critical to human immune system function (e.g., Balter, 2017). These schizophrenia polygenic risk regions have also been found to be associated with creativity (e.g., MacCabe et al., 2018; Power et al., 2015; Srinivasan et al., 2016).

Despite these caveats and warnings, it cannot be said there were *no* fitness advantages of some types of current psychopathologies in the ancestral environment. An important concept in this discussion is *evolutionary stable strategies*, or ESS. Evolutionary psychologists emphasize that natural selection typically favors a single predominant behavioral pattern in the population, yet alternative or dissimilar patterns may develop and persist. The dominant behavior is said to have *high penetrance*, with the highest level being called "fixed" in the population. These patterns (both the dominant and alternatives) are labeled ESS. Investigators further purport that an optimal and stable ratio develops between the dominant and alternative behavioral patterns. For example, in ant and bee colonies, a consistent ratio of fertile insects to sterile insects develops and stabilizes over time. If a population of insects becomes dominated by the sterile members, it will create an inherently unstable structure and will more than likely become extinct. British biologist John Maynard Smith coined the term *frequency-dependent selection*, to describe the phenomenon wherein selection occurs when relative fitness depends on the frequencies of genotypes. In negative frequency-dependent selection, the fitness value of a trait will decrease as its frequency in the population increases. In practice, this means that natural selection places an upper limit on a trait's penetrance in the population, which then allows for an alternative and usually dissimilar behavioral trait. However, negative frequency-dependent selection does not mean any trait will decrease as its frequency in the population increases, but only those traits whose "fitness" varies inversely with its population frequency.

Interestingly, a recent study (Bolnick & Stutz, 2017) noted that evolutionary lineages may diverge by two different processes. The first is that varying environments may favor different phenotypes, resulting in increasing differences between or among different populations. The second may be less common, and it is an example of negative frequency-dependent selection, whereby rare phenotypic variations may sometimes be favored over more common ones, which leads to greater diversity *within* a population. Bolnick and Stutz tested the latter process in fish in the wild. They found evidence that negative frequency-dependent selection appeared to prevent the fixation of a single genotype (and thus overly predominant phenotypes), because it retained the lesser adapted and rarer alleles at a low equilibrium frequency. They noted there might have been some unspecified benefit of these rarer phenotypes, which apparently outweighed their overall fitness liabilities. One possible benefit is the protection of the overall species by preserving a rarer but related species or subspecies. One example might be the present-day Cavendish banana, which is the economically dominant species worldwide, but it is currently threatened by a fungal disease, Fusarium wilt. Because of the maintenance of rarer but related banana species, the world may lose the Cavendish banana, but it might gain other banana types. With these evolutionary foundations established and within these disparate contexts, I shall begin my discussion with a group of psychopathologies known as personality disorders that are at least 10 times more prevalent than schizophrenia (schizophrenia has a 1% worldwide prevalence) and are also as cross-cultural and universal (Coolidge & Segal, 1998; Segal, Coolidge, & Rosowsky, 2006).

Personality Disorders

A *personality disorder* is currently defined as "an enduring pattern of inner experience and behavior that deviates markedly from expectations of the individual's culture, is pervasive and inflexible, has an onset in adolescence or early adulthood, is stable over time, and leads to distress or impairment" (American Psychiatric Association, 2013, p. 645). The history of personality disorders can be traced as far back as the fourth century BC to a contemporary of Aristotle, Theophrastus. In his book *Characters,* written in about 320 BC, he described 30 kinds of people in Athenian society, and many of his caricatures mirror modern personality disorder diagnoses. In more recent history, personality disorders have also been called *character disorders,* which predates modern personality disorder diagnostic systems, and that name was no doubt influenced by Theophrastus' book. Currently, there are 10 personality disorders listed in the *Diagnostic and Statistical Manual of Mental Disorders* (5th ed.) (*DSM-5*), although 21 different personality disorders have appeared in the seven versions of the *DSM* since 1952. Some of the more common or better-known personality disorders that appear in the general literature are antisocial, borderline, narcissistic, paranoid, obsessive-compulsive, histrionic, passive-aggressive, and schizoid. Personality disorders are known to be highly heritable due to a confluence of genes (polygenic; e.g., Torgerson, 2000, 2009 Torgersen et al., 2000) and manifest early in life, which is also evidence they are under tight genetic control (e.g., Coolidge, Thede, & Jang, 2001; Robins, 1966). Torgersen (2009) found that the overall influence of genes for personality disorders is about 40 to 50%. It is thought that the genes that influence personality disorders are the same ones that influence normal personality dimensions, such as extroversion and introversion, except that the phenotypes that result have gone awry in some regard (e.g., Segal et al., 2006). For example, the extreme phenotype for extraversion might be the histrionic personality disorder, and the extreme phenotype for introversion might be the schizoid or avoidant personality disorders.

One evolutionary perspective on the ultimate origins of personality disorders is this seemingly disparate group of disorders might have developed in the ancestral environment because of the varying ESS that would be employed to successfully "navigate" status hierarchies (which are characteristic of nearly all primates). A status hierarchy implies that animals living in groups fall along a continuum from dominant to submissive, based on size, aggressiveness, intelligence, and other characteristics. By evaluating the success of fighting or submitting or working to gain allies, individuals maximize their chances for obtaining mates, food, or other resources. It is thought that there are some inherent genetic predispositions (i.e., temperaments) that drive people to their "natural" place in a social hierarchy. For example, inheriting an above-average tendency for aggression and a below-average tendency to avoid harm might place an animal toward the dominant end of the social hierarchy. Those with the exact opposite tendencies would be closer to the submissive end of a hierarchy. As noted earlier, by not challenging every other animal for a place in the hierarchy, organisms are able to maximize their resources, in part by not wasting them on unnecessary competition.

Antisocial Personality Disorder

The personality disorder that has historically received the most attention is the antisocial personality disorder (also known as psychopathy or sociopathy). Its essential feature is a callous disregard for the rights or feelings of others. In a classic longitudinal study of children with conduct disorders, Robins (1966) concluded that biology (genes) must play a strong role in the development of antisocial personality disorder, as one of the strongest predictors of adult antisocial personality disorder behavior was an antisocial father, regardless of whether the biological father was present or absent in the child's early life. Antisocial personality disorder is also much more prevalent in men (3%) than in women (1%), which likely accounts for Robins' observation that the father's genetic influence is powerful. Although prevalence rates for antisocial personality disorder in prisons often reach 40% (e.g., Comer, 2015), its true prevalence is likely lower, with a minimum prevalence estimated at 15% (e.g., Coolidge, Segal, Klebe, Cahill, & Whitcomb, 2009). This discrepancy in prevalence rates is likely due to broader definitions of antisocial behavior, including dissocial behaviors, which often lack the callousness and pervasive irresponsibility of antisocial adults. Antisocial personality disorder also has an early onset, often by age 7 years (e.g., Robins, 1966), although the diagnosis is not typically applied until later adolescence or early adulthood. The childhood or adolescent onset is usually labeled *conduct disorder*, and it is important to note that about half of children who meet the criteria for conduct disorder will not be later diagnosed as antisocial(e.g., Coolidge, Merwin, Wooley, & Hyman, 1990; Robins, 1966).

The *DSM-5* lists the major feature of antisocial personality disorder as a disregard for the rights and feelings of others. Specific symptoms include failing to adhere to social norms regardless of the nature of one's society, deceitfulness and lying, impulsivity and failing to plan ahead, total disregard for one's own safety or the safety of others, irresponsibility (with family members, others, work, school), and a lack of genuine remorse.

Evolutionary psychologists often focus on differing reproductive strategies between men and women as just one of the ultimate causes for antisocial behavior. It is abundantly obvious that a man can produce more offspring than a woman, with anecdotal reports of men having hundreds of children (like Genghis Khan), while a woman in Russia from the 1700s apparently had 69 children. Clearly, a man can produce offspring, which carry his genes, without having any responsibility for their upbringing. Even if he abandons the mother of his children after conception, some of these children will reach reproductive age. However, according to evolutionary psychologists, a woman's best reproductive strategy is different. Her genes will enter the gene pool only if her children are cared for and reach reproductive age. While there might be some minimal selective advantage for a woman to be promiscuous (e.g., self-promotion, obtaining resources, etc.), there is a raft of empirical and theoretical evidence that a man has more to gain from being promiscuous.

Sex: A Provisioning Strategy

If all men adopted the "promiscuity" ESS, this trait would soon lose its evolutionary value, because fitness is always determined relative to alternative strategies and traits. If most men stayed around after the conception and birth of their offspring, and, further, if they provided for the woman and children, their genes would have a higher likelihood of entering the gene pool than the genes of a promiscuous ESS. Evolutionary psychologists posit that women may have used sex as a means of keeping men around for the provisioning of offspring. However, if a man knew exactly when a woman's small window of fertility occurred (about 48 hours a month), he might only pay attention to women a few days a month. Thus, it is suspected that women who did not physically display their brief period of fertility were at an advantage and were more likely to keep men around for the provisioning and protection of their children. In many mammals, females of the species typically have obvious changes in physical and sexual behavior just prior to ovulation during the highest period of fertility. It is most dramatic in extant chimpanzees, humans' closest genetic relatives: There is very discernable genital swelling in female chimpanzees when they are ovulating. However, while it may be true that a woman in estrus does not know exactly when she can conceive, it is provocative to ask whether a woman's highest fertility period is really completely hidden.

Is a Woman's Estrus Cycle Really Hidden?

Although historically the standard mantra in evolutionary psychology is that a woman's period of estrus is hidden (even from herself), in a now classic study of lap dancers (partially clothed women who dance in close approximation to their clients), it was found that dancers' tips were significantly greater (by almost 30%) on high-fertility days than on low-fertility days (Miller, Tybur, & Jordan, 2007). Of course, this finding raises a number of issues, including determining the extent to which men and women mutually interact to produce this finding and the extent to which the behavioral changes are under conscious or unconscious control. For example, there is evidence that women choose more sexually attractive clothing during their high-fertility days, their body odor is perceived as more attractive by men during high-fertility days, and men are more jealous of their partners during high-fertility days (Haselton & Gildersleeve, 2011). A signaling hypothesis has been proposed, which states that women may benefit from actively signaling their fertility (Thornhill & Gangestad, 2008). Another explanation is based on the standard mantra that women conceal their estrus to keep men around for provisioning even through non-fertile days; however, their concealment appears "leaky," that is, fertility cues still manage to enhance fitness in spite of largely being hidden. For example, women may give off cues to their fertility to compete with other women and to gain men's attention (Haselton & Gildersleeve, 2011). The latter notion (gaining or extending male attention) is related to the concept of male parental investment.

Parental Investment

The investment in the child's development is called *parental investment*, and thus evolutionary psychologists speak of male parental investment (MPI) when they describe this tendency for the male to "stick around" and provide resources for the raising of the child. It is also thought that a woman will evaluate a man for his "wealth" (or his ability to obtain resources), that is, a man with a "high MPI." It might be predicted, then, that women will only be attracted to "good dad" types and shun the "bad boy" or the "rebel," but this appears not to be the case. While women more often appear to marry stable (or semi-stable) men, bad boys like the movie star James Dean become mythical attractive figures in American culture. Interestingly, there is some empirical evidence that when they are ovulating or choosing to have an affair, women have an elevated predilection for bad boy types, particularly if their partner is not the bad boy type (Durante, Griskevicius, Simpson, Cantu, & Li, 2012). Furthermore, after women marry these supposedly stable men, that is, men with a high MPI, some men subsequently turn into bad boys by philandering (infidelity rates for married men range from about 20 to 75%). However, the question might be asked, do they become bad boys, or were they always bad boys and just pretending to be high in MPI?

Most evolutionary psychologists think the latter is true. They surmise that most men unconsciously fool their potential wives into thinking they are high in MPI, and after they have fulfilled their biological imperative (have a child or two), they often divorce their wives and move on (although there is certainly great within-group variation). Divorce statistics support this hypothesis, and more often these men then choose a younger wife, that is, one who can provide him with the promise to fulfill his biological destiny (i.e., more children). According to evolutionary psychologists, the phenomenon of men marrying younger women (and conversely women marrying older men) mutually fulfills our biological predispositions: A man marries younger because younger women have more childbearing years ahead than older women, and women are attracted to older men because they have had more time to collect resources (wealth), demonstrate their ability to achieve wealth (through power), and show their wisdom, stability, and dependability.

Overall, and despite the previous hypotheses, it appears that women tend to favor men who will make good dads. However, we still have not accounted for any tendency of women to secretly or openly favor "bad boys." From an evolutionary perspective, it may be speculated that some bad boys will be successful in reproducing because of their strength or guile, and the women who fall prey to these alternate strategies (other than investment of resources) might have sons who resemble their fathers. The sons may employ the same alternative bad boy strategy, and thus the genes perpetuating these behaviors survive.

In this "bad boy" context, antisocial behavior in the ancestral environment would have fulfilled the same goals as it does in present society. According to evolutionary psychologists, some men "fool" women into sex by making them think they are wealthy and will make good dads. In the ancestral environment, the reproductive conditions were similar: Eggs were precious, sperm was cheap. Interestingly, according to anthropologists, present-day sexual dimorphism (i.e., men tend to be bigger, heavier, and taller than women) is an evolutionary remnant of the time when there was competition among men for women. In contemporary society, those diagnosed with antisocial personality disorder tend to move from place to place, which supports their promiscuous behavior. However, in ancestral environments (called the *environment of evolutionary adaptedness* [EEA]), it might have been more difficult to move about the landscape without the protection of others. Nonetheless, it is suspected that group size in more recent EEAs like *Homo heidelbergensis* about 800,000 years ago, was large enough (e.g., 150) to support significant promiscuity without the dangers of moving about the landscape. Thus, EEAs for various psychopathologies might differ from the remote past because of persistent factor spanning million of years to the more recent past.

There may also have been some "positive" features of antisocial behavior in the ancestral past besides the obvious "negative" ones, like lying and irresponsibility. Antisocial types would probably make for rather fearless warriors, with their impulsivity, disregard for their own safety, and lack of remorse. Thus, while obviously maladaptive in present society, antisocial behavior in its EEA may have had sufficient reproductive value to persist until the present day.

Paranoid Personality Disorder

According to *DSM-5*, people with paranoid personality disorder are pervasively mistrustful and suspicious of others. Like the antisocial personality disorder, paranoid personality disorder is largely under genetic control, has a similar prevalence rate (2–4%), and is seen more often in men than in women. Its symptoms include a suspicion that other people are trying to harm, exploit, or hurt one; a preoccupation with and suspicion of the faithfulness, loyalty, or trustworthiness of others, including spouses, family members, and friends; reluctance to confide in others; readily interpreting malevolent meanings and motives in other people's behaviors; bearing grudges and being unforgiving; and quickness to counterattack perceived threats.

Again, similar to antisocial personality disorder, the higher rates of paranoid personality disorder in men may be explained by differing reproductive strategies between men and women. Both in the present environment and in the EEA for this disorder, what woman has ever wondered whether a baby, especially while in her womb, is really hers? None! What that means is that a woman has absolute biological certainty of her share of the child's genes. Men, however, never come close to a woman's parentage guarantee. The man must take a woman's word that the child is his, and from his biological perspective, this is a perilous position. If he makes an inordinate parental investment, that is, he decides to fertilize only one woman's eggs, and she is impregnated by another man, then his genes do not enter the gene pool. Thus, suspicions and doubts of the faithfulness of a spouse would be more biologically useful for a man than for a woman. However, some women also routinely question their spouse's faithfulness. Why would a woman care if a man shares his virtually limitless sperm with other women? According to contemporary empirical

evidence, a woman cares more about her spouse sharing resources, for example, money and time, than sharing sex because it might result in fewer
resources for her own children. Even an emotional commitment to another
woman might be troubling since it could mean losing other kinds of support,
like emotional support and time commitments. In the language of evolutionary psychology, it is not the number of offspring produced that is important, but the number of offspring who reproduce.

With regard to other paranoid personality disorder symptoms, in both
the EEA and present environments, it might be useful to have some, but not
all, members of a group genetically predisposed to suspecting harm, exploitation, or deception. As noted in Chapter 6, the failure to detect a pattern
or event (false negative) can ultimately be much more harmful (fitness reducing) than detecting a pattern that does not really constitute a real pattern
or event (false positive). Especially in the EEA, imagine the consequences of
failing to see an oncoming predator versus seeing a predator that really is not
there. Although the latter behavior might be deemed foolish or wasteful, the
consequences of the former predisposition could be deadly. Even in today's
environment, there are still harmful behaviors that are associated with false
positives, like allergies and wasting one's energies on too many false alarms.
This phenomenon may also help to explain the tendency of some to believe
whole-heartedly in conspiracy theories, no matter how outlandish the conspiracy. From an evolutionary perspective, it may be as if those predisposed
to conspiracy theories were in essence saying, "Yes, it's outlandish, but at least
I won't be surprised like you if it turns out to be true." Further, a comparison
of the severity of consequences between false positives and false negatives
hinges on the nature of the threat.

As noted in Chapter 6, the tendency of modern humans to see patterns
that are not real or to see meaning in events that are not really meaningful
is called *apophenia*. A subcategory of apophenia is *pareidolia*, which is the
tendency for humans to perceive a face or pattern where one does not exist.
Ontogenetically (the genetics of early development), it would be important
for infants to recognize and distinguish faces for sustenance and sociality.
Astronomer Carl Sagan, in his 1996 book, *The Demon Haunted World*, noted
that infants who could not recognize their parents' faces were less likely to
win their parents' hearts and thus less likely to prosper. Evolutionarily, it
would be critically important to recognize faces (and other objects, both animate and inanimate) immediately for those two reasons, and to recognize
predators and other threats in the environment. Sagan informally labeled

the "inadvertent side effect" of this phenomenon "the pattern-recognition machinery" (p. 45), where "we sometimes see faces where there are none" (p. 45). Sagan also gave examples of seeing faces in rocks, vegetables, and wood, and even the face of Jesus, which has been seen in tortillas, windows, and other objects. Skeptic and writer Michael Shermer (2011) has claimed that brains are "belief engines" with strong predilections to detect a pattern where no pattern might exist and attribute some sense of agency or causation to the patterns rather than the mundane alternative of simple chance or randomness. Some also suspect that the false-positive bias may have been a perceptual design flaw, which resulted in reducing false negatives, particularly under threatening conditions (e.g., Dodgson & Gordon, 2009).

Borderline Personality Disorder

People diagnosed with borderline personality disorder live highly unstable lives: Their relationships are rarely smooth and often very rocky, and even their views of themselves are unclear, distorted, or confused. They are highly impulsive in their actions, which can be manifested in gambling problems, binge eating, drug and sexual addictions, speeding or driving recklessly, or sudden and hasty decisions with major consequences (like injury to self or others). Psychologist Theodore Millon and his colleagues (Millon, Millon, Meagher, Grossman, & Ramnath, 2012) claim that a core feature of the borderline personality disorder is affective (emotional) instability, particularly regular bouts of intense anger directed at other people, including family members. Other symptoms include extreme efforts to avoid abandonment (real or imagined), recurrent suicidal threats, self-mutilating behavior, strong feelings of emptiness, and transient paranoia or dissociative behaviors. Borderline personality disorder is also under strict genetic control and usually starts by adolescence or earlier. If it does arise in childhood, it may manifest as an attention-deficit/hyperactivity disorder (ADHD) or a mood disturbance like bipolar disorder. It has a higher prevalence rate than most other personality disorders (2 to 6%), and it is seen about three times more often in women than in men.

On the surface, it would appear that there may have been little or no adaptive value to most of the symptoms of borderline personality disorder in the EEA. However, one possibility for a cluster of the borderline symptoms may again reside the ubiquity of social-dominance hierarchies in all extant primate species, which makes it highly likely that social hierarchies were prominent in ancient hominins. Studies of modern hunter-gatherers and cross-cultural studies support this supposition as well. Subdominant members of a group may have engaged in suicidal gestures and self-mutilation to remove themselves from any perceived competition with higher ranking members. Indeed, it has long been observed in extant primates that abnormally acting members are often ignored, and primates who no longer defend themselves are not subjected to further abuse or aggression (e.g., Bernstein & Gordon, 1974). These self-injurious behaviors may have beneficial, fitness-enhancing secondary effects. Dellinger-Ness and Handler (2006) have noted that a survey of animal research shows that self-injurious behavior nearly always follows stressful situations, and further, self-injurious behavior appears to return animals' heart rates to baseline levels after the experimental introduction of stress. Also, as anger and aggression levels in humans are under strong genetic control, those who have inherited higher propensities for anger and aggression may frequently engage in self-injury in order to harm themselves before others do *and* to relieve the stress of becoming intensely angry (the latter phenomenon in Freudian psychoanalysis is called *catharsis*).

Another prominent symptom of borderline personality disorder is an identity disturbance, in which patients have unstable self-images or a poor sense of self. Again, this symptom in the EEA may be related to the social-dominance hierarchy, where a strong sense of self might be more closely associated with challenges, anger, and aggression from others. Those with an unstable or even flexible self-image might be less likely to encounter opposition in a group's hierarchy.

Although the adage "look before you leap" might well have been safer in a majority of situations in the EEA, the impulsivity of borderline personality–disordered hominins might also have had occasional advantages. Impulsively eating unknown substances in times of starvation might well have led to nutritious discoveries, although admittedly it might just as often removed an individual from the gene pool. As has been said tongue-in-cheek, all mushrooms are edible at least once. Additionally, there may have been occasional advantages to impulsive environmental exploration (e.g., discovering novel water and food sources) and even impulsive sexual explorations with other hominin group members. Certainly, the evidence for the limited Neandertal genetic introgression into *Homo sapiens* suggests that both human types must have had some brave, intrepid, or impulsive members as the more robust (shorter and stockier) Neandertals would have appeared rather differently to the gracile (taller and thinner) *Homo sapiens* and vice versa (Harris & Nielsen, 2016; McCoy, Wakefield, & Akey, 2017; Overmann & Coolidge, 2013).

Schizoid and Avoidant Personality Disorders

On their surface, these two personality disorders are behaviorally similar: Individuals with these disorders avoid interactions with other people. Millon long ago proposed that schizoid individuals *passively* avoid others while avoidant individuals *actively* avoid others. I respectfully disagree with his opinion (sadly, he passed away in 2014). I think both types actively and passively avoid others, as I shall explain shortly.

Schizoid personality disorder is frequently misunderstood. The word *schizoid* is virtually synonymous with the word *loner*. People with schizoid personality disorder neither desire nor enjoy having relationships with anyone, even family members. They highly prefer their own company as well as activities and hobbies that can be conducted alone, like stamp collecting. To other people they seem cold, emotionless, and lack affect (emotions like joy, sadness, delight, etc.). Further, they seem immune or indifferent to praise or criticism. Like all personality disorders, it is highly heritable, polygenic, in this disorder more prevalent among men than women. Schizoid personality disorder also has an early onset, often evident in early childhood; such children or adolescents appear to be friendless. Although the *DSM-5* states that personality disorders may manifest in the cognitive domain, there are no well-known or well-established cognitive deficits associated with the schizoid personality disorder. People with schizoid personality disorder simply appear to be loners and lack any desire to interact with anyone else. Their remoteness also means that the prevalence of schizoid personality disorder is difficult to assess. Those with the disorder will not voluntarily seek out mental health treatment or any kind of psychotherapy. thus they are not often seen in clinical settings.

While behaviorally similar, individuals with avoidant personality disorder at least desire interactions with other people. However, they have strong aversions to social interactions, often feeling highly inadequate. They are hypersensitive to imagined or perceived slights, criticism, rejection, or negative evaluations from others. Thus they shy away from social activities and interpersonal interactions. When confronted with interacting with others or speaking in front of others, they are very restrained and inhibited. When probed clinically, they readily admit that they feel inept, inadequate, awkward, and unattractive compared with their peers. The avoidant personality disorder is also highly heritable and polygenic and appears equally among men and women. Finally, it is important to note that personality disorders may be conceptualized as confluences of various normal personality traits gone awry.

Once again, one of the most likely scenarios for the persistence of these two personality disorders may be related to the social-dominance hierarchies. Behaviors that remove oneself from the vagaries and stresses of forming and maintaining alliances within subgroups of a larger group must have had some evolutionary advantage, such as being less likely to be subject to aggression and abuse from higher-statured individuals. If all members of a group adopted this ESS, the group would undoubtedly probably die out, owing to their reduced reproductive activities, especially compared to more socially interactive groups.

Another evolutionary advantage of socially resistant individuals might be related to communicable diseases. It is possible that in times of plague or socially transmittable diseases, the genes of the socially resistant might survive because of their hosts' reduced likelihood of contracting diseases transmitted through social interactions. Interestingly, in the Bolnick and Stutz (2017) study mentioned earlier, negative frequency-dependent selection often appears as a mechanism for the maintenance of genes that code for major histocompatibility complex proteins. Thus, one benefit of the rarity of those genes may be related to regulation of a healthy immune system. I shall return to this argument shortly.

Schizotypal Personality Disorder and Schizophrenia

Although individuals with schizotypal personality disorder are socially awkward and have great social anxiety, they also tend to have subtle cognitive impairments, particularly with regard to perceptual distortions and unusually loose or "weird" thinking. It has been suggested that this disorder might be on a continuum with schizophrenia, a major psychotic disorder in which hallucinations and delusions figure prominently. The common symptoms of schizotypal personality disorder are making meaningful associations with various environmental stimuli that are not meaningful and usually random (i.e., apophenia), having inflated ideas of reference (e.g., "famous people are aware of me"), having very odd or magical beliefs that are inconsistent with one's culture or subculture, experiencing bodily illusions (ability to float, etc.), exhibiting highly superstitious behaviors, and having beliefs in telepathy, clairvoyance, or a "sixth sense," such as curing cancer by the laying on of the hands. To others, individuals with schizotypal personality disorder appear very odd, eccentric, and emotionally constricted. Others find their speech strange or weird, excessively metaphysical or metaphorical ("The wind is in the meadows" might be a reply to "How are you?"), and idiosyncratic (always replying "Good morning" in the afternoon), with digressive ideas in speech, yet without actually being completely incoherent or senseless. The schizotypal personality disorder is highly heritable, polygenic, and diagnosed in men slightly more than in women. Some theorists view schizotypal personality disorder as an "arrested" subtype of schizophrenia or a version of schizophrenia that did not develop fully, yet both disorders may be measured along a psychotic continuum.

Schizophrenia is considered a major psychotic disorder, and its two chief symptoms are hallucinations and delusions. The hallucinations may occur in any of the five senses but the most common are auditory hallucinations, in which individuals report hearing voices inside their heads. Analyses of the voices' content tends to reveal that they are mostly mundane, like "brush your teeth," although in a very small percentage of people with schizophrenia the voices urge killing other people and sometimes their own children or other family members. *Delusions* are defined as false beliefs, and the more common delusions tend to be persecutory, that is, someone or some group (e.g., FBI, CIA) is "out to get them." When I was a forensic psychologist in a state mental hospital, the other two most common delusions among schizophrenic patients were nihilistic and grandiose. In nihilistic delusions, patients claimed they were worthless and nonexistent. Patients with grandiose delusions often claimed to be Jesus Christ). A casual observer of a ward of schizophrenic patients might also sense a kind of hyper-religiosity, and not just among the patients who thought they were Jesus Christ but among other patients as well (when I served as a clinical psychologist at a federal mental hospital in southern India, Hindu schizophrenic patients often claimed to be one of the Hindu gods, and often it was one of their family's traditional gods).

Another common observation made about schizophrenic patients is their creativity. However, this association is often overstated, in my opinion, as schizophrenia is very debilitating, making a person unable to work, go to school, or even take basic care of themselves (e.g., washing, bathing, eating properly). In addition, a strong majority of schizophrenic patients are not creative at all. I once had a patient who stuck a needle in his eye to see "God" and another who impaled his testicle with a pencil to be seen at the medical clinic. My reservations aside, and as noted earlier, it is suspected that schizophrenia and creativity have common but complex genetic roots (e.g., MacCabe et al., 2018; Power et al., 2015; Srinivasan et al., 2016).

Schizophrenia is not considered a personality disorder, although schizophrenic patients often have comorbid personality disorders, such as antisocial and schizoid personality disorders. Schizophrenia, like personality disorders, (1) is highly heritable, (2) is polygenic (small effects from a multitude of genes), (3) has no cure, and (4) is difficult to treat even with drugs and psychotherapy. In individuals who have been diagnosed with schizophrenia, the likelihood of developing schizophrenic symptoms again after a period of remission is about 50%. Its prevalence rate worldwide is about 1%, and it is known to occur in nearly every culture. The prevalence rate tends not to vary much even with catastrophic conditions like war, terroristic acts, famine, or pestilence, which attests to its biological nature and that it is under strong genetic control.

Psychologist Paul Meehl (1962), in a classic paper that has weathered well, claimed that one must inherit the genetic predisposition to develop schizophrenia, which he called *schizotaxia*. He theorized that even under dire, threatening environmental conditions or unknown stressors, an individual could *not* develop schizophrenia unless they were schizotaxic. Also, if an individual was not schizotaxic, then that person could *not* develop schizophrenia under any circumstances, although they might temporarily exhibit some schizophrenic-like symptoms, like being under the influence of hallucinogenic drugs. Further, he hypothesized that if an individual was schizotaxic, they might not develop schizophrenia (although they could remain a genetic carrier). These bold hypotheses are still valid well over 50 years later, although there is no genetic test yet developed to assess schizotaxia. Psychologists have since expanded Meehl's hypothesis to other disorders in what is called the *diathesis-stress hypothesis*, which states that a genetic predisposition (diathesis) interacts with one's environment (stress) to produce a disorder.

Recent genomic studies suggest that schizophrenia is a complex, genetically based condition of at least 108 interacting genes (Schizophrenia Working Group of the Psychiatric Genomics Consortium, 2014, in a study of 37,000 cases of schizophrenia and 113,000 controls). This array of genes ultimately creates a "family" of many schizophrenic types (the latter is Meehl's concept of schizotypy) with varying comorbid personality disorders and other psychopathological symptoms.

The Evolutionary Paradox of Schizophrenia

The evolutionary paradox of schizophrenia is that it is a highly heritable disease, is highly debilitating, has an early onset (in adolescence of young adulthood), and has negative fitness effects for a strong majority of its members. However, it persists worldwide and cross-culturally. Crespi, Summers, and Dorus (2007) studied 76 gene loci known to be associated with schizophrenia and found positive selection for 28 of the genes in regions also associated with creativity, imagination, and divergent thinking. They proposed that the genetic susceptibility for schizophrenia may have been a secondary result of the positive selection for creative cognitive traits. Indeed, British psychiatrist Timothy Crow (1997) has long argued that schizophrenia is the price that *Homo sapiens* paid for their sophisticated language skills. Crow hypothesizes that the cerebral hemispheres developed their relative independence and differential functions only in the last 250,000 years or so. He believes that this relative independence was critical to language's generativity. However, he proposes that schizophrenia may reflect a breakdown of the bilateral coordination that normally exists between the two hemispheres regarding its language functions. In this fashion, psychotic thinking arises because of the this breakdown in hemispheric specialization and lack of hemispheric coordination, although Crow does not specify any positive selection benefits of this failure. The Schizophrenia Working Group of the Psychiatric Genomics Consortium (2014) also showed that some of the genetic regions for schizophrenia code proteins related to the major histocompatibility complex, that is, proteins that recognize foreign or alien molecules and are responsible for a healthy immune system. Further, there is little empirical evidence for Crow's hypothesis, although it remains provocative, as it is difficult to argue against the notion that one must learn of Jesuits (by way of language) to have delusions about them. [Note: Jesuits have long been a topic of mystery and intrigue (if not outright delusions) since their founding in 1534]. Nevertheless, it seems difficult, if not impossible, to develop delusions (false beliefs) without language, whereas hallucinations (i.e., perception in the absence of stimulation) seem possible without language.

In addition to healthier immune systems, I favor the hypotheses that the positive selection benefits of schizophrenia may reside in a small minority of its members who can think creatively and divergently. It is also possible that the positive selection benefits reside in family members of schizophrenic patients who carry similar but not identical genes. In this way, related family members may carry the risk for schizophrenia (schizotaxia) but are not greatly debilitated by the disease and still benefit by being able to think more creatively and imaginatively.

Stressed throughout this book is the notion that novel problem-solving and creative solutions may have been the key to the survival and ascendancy of modern *Homo sapiens*. It is also important to reemphasize that the onset of schizophrenia is in adolescence or early adulthood, and slightly earlier for men than for women. In the EEA, this may have meant that humans who inherited schizophrenia did so after they had reproduced. Thus, the onset of a debilitating psychosis before reproduction would be naturally selected against, and the age-of-onset statistics support this hypothesis. Of course, this would mean that schizophrenia may not have been selected for its adaptive value, and its onset prior to reproduction would be selected against, which is exactly what is found in schizophrenia prevalence rates. For a provocative view that argues *against* the idea that mental diseases are adaptations in their EEA, see Dubrovsky (2002).

Despite Dubrovsky's argument, the evidence for positive selection benefits of schizophrenia promoting inclusive fitness regarding some distinctly human traits is appealing. When schizophrenia starts after adolescence, language learning, social rules, regulations, and skills are all well in place. The debilitation at this period does not "de-program" language, as many schizophrenics can report their hallucinations and delusions in great detail. Further, grandiose hallucinations and delusions are common, and schizophrenic patients will report they are somebody more important (like Jesus Christ or an apostle) or are of much greater influence than they really are. Also, as noted previously, schizophrenic patients often tend to be hyper-religious, and this effect is cross-cultural. Is it possible that belief in a "higher power" and "hyper-spirituality" are positive selection traits?

Evolutionary psychologists have speculated that religions have strongly promoted human inclusive fitness by providing a set of rules and myths for guiding themselves to overcome their genetic self-interests. It may be that schizophrenia and family members of schizophrenic people helped promulgate the creative and inventive myths of religions. Furthermore, since religious figures are, by definition, translators and intermediaries of the unknowable and the unfathomable, some schizophrenic people might be able to assist in such matters. Psychiatrist R. D. Laing (1967) took the radical standpoint that that schizophrenic patients were more in contact with their inner world than "brainwashed" normal people were. Laing wrote, "Let us call schizophrenia a successful attempt not to adapt to pseudo-social realities" (p. 58). I do not fully agree with Laing's stance in this quote, as Carl Jung even recognized that often one's inner world may overwhelm one's conscious mind to the point of incoherence. I do believe that people with schizophrenia or the relatives of schizophrenic people may have increased inclusive fitness by their creative contributions to religion. It is interesting and provocative to note that in contemporary Western societies, virtually the only group of people that regularly admit they have visions and contacts with God are religious people and psychotics. Yet, what theist could state with absolute certainty that God could not talk to people or schizophrenic people? It may have been that schizophrenic people in the EEA were the first intermediaries between humans and unknowable and unfathomable mysteries (like shamans). Atheists often argue that since there is no god, notions of gods are delusional and visions of gods mere hallucinations. Here, I would note that it does not matter whether shamans or god intermediaries could contact the "beyond." It appears sufficient that people *thought* they had access to the divine. In summary, the social rules and myths afforded by supernatural belief systems may have promoted inclusive fitness by helping humans overcome their genetic self-interests (see Silverman, 1967, for a classic paper of the shaman hypothesis and schizophrenia; also see Fiala & Coolidge, 2018; Singh, 2018).

Attention-Deficit/Hyperactivity Disorder

One of the most prominent psychological issues of childhood and adolescence is attention-deficit/hyperactivity disorder (ADHD). It was formerly labeled *minimal brain dysfunction* (MBD) and *attention deficit disorder* (ADD). In a review of 102 studies of 171,756 children worldwide, Polanczyk, de Lima, Horta, Biederman, and Rohde (2007) found a prevalence of ADHD of 5.3%. In adults, ADHD occurs at about half the child/adolescence prevalence rate. It is diagnosed more frequently in boys than in girls at a ratio of 4 or 2:1. The primary symptom of ADHD, as the name is meant to imply, is a problem paying attention, including an inability to sustain attention to a task, lack of attention to details, failure to follow instructions, difficulty prioritizing and completing tasks, poor organization, and poor planning. About half of the children or adolescents who have prominent attention deficits have comorbid (problems occurring at the same time) hyperactivity and impulsivity, including an inability to sit still, excessive fidgeting, interrupting others, difficulty awaiting one's turn, excessive restlessness, and sleep problems. Even a cursory review of all of these symptoms might suggest that nearly all children might qualify as ADHD at some point in their development, and not surprisingly, many of the estimated prevalence rates for ADHD are thought to be overinflated. In prison populations (adolescent and adults), the prevalence rate of ADHD increases dramatically to about 15% or more (e.g., Coolidge et al., 2009). This finding suggests that there is a comorbidity of ADHD and antisocial behavior, although its exact nature remains elusive. It is also known that ADHD is highly heritable and polygenic, with a few specific genes having been identified, particularly ones associated with the production of dopamine.

However, if a 5% prevalence rate is taken as a conservative estimate of ADHD, then why the persistence of this psychological problem in adults that leads to significantly higher rates of divorce, unemployment, educational problems, suicide, and comorbid mental problems? Jensen et al. (1997) were among the first to suggest that ADHD may have been adaptive in the EEA. Their initial reasoning, as has been stressed throughout this book, is that movement is critical to an organism's success in its environment, both to acquire resources and to avoid threats. Animal studies have long shown that increases of time in activities, foraging time, and speed of locomotion predict successful resource acquisition. Further, foraging speed has been shown to be positively correlated with diet quality in over 40 extant nonhuman primate species (DeLouize, 2017). Coupled with appropriate motor activity, the ability to rapidly shift one's attention might be of great value in a novel or threatening environment. With regard to the symptom of impulsivity, a quick response might be favored in situations where delay might have more dastardly repercussions. Jensen et al. labeled such individuals "response-ready," as opposed to more phlegmatically (slow, relaxed) predisposed individuals, whom they labeled "environmentally challenged." They reasoned that response-ready individuals might have been more successful in warfare or under more challenging or difficult environmental conditions. One final and important comorbid symptom with ADHD is insomnia. Some studies have shown that up to 75% of people with ADHD have trouble falling asleep or staying asleep, have restless sleep, or have all of these symptoms. At the time of the tree-to-ground-life transition from *Australopithecus afarensis* to *Homo erectus* about 1.9 million years ago, it is strongly suspected that group sizes increased dramatically. Imagine the value of a few individuals in a large group who did not sleep well initially or could not sleep through the night. Once again, these response-ready individuals would be more alert to dangers and predators in the surrounding environment, and they could serve as human "alarm clocks" by alerting more soundly sleeping group members to immediate threats.

Eating Disorders: Anorexia Nervosa and Obesity

Eating disorders (eating too little or too much) are common throughout the world. One of the most baffling medical disorders and deadliest of the eating disorders is anorexia nervosa (literally, loss of appetite because of nerves). Estimates of its mortality rate ranges from 5 to 20%. Its symptoms are a severe loss of weight (more than 15% or more under an ideal weight), fears of gaining weight, abnormal body self-perceptions, and hyperactivity or excessive exercise. It affects women more frequently than men, and in women there is usually a cessation of the menses. Nesse (1984) offered an evolutionary basis for anorexia in which he hypothesized that the consistency of the symptoms cross-culturally suggests an adaptive mechanism related to swings in food availability, from feast to famine. Guisinger (2003) has remarked that evolutionary biologists propose that the cessation of menses might be a kind of evolutionary protective device against reproduction in dire times. However, as Guisinger notes, that explanation in understanding the hyperactivity and intense exercise that accompanies anorexia or why it sometimes occurs in adolescent and adult men. She proposed an "adapted to flee famine hypothesis" (AFFH; p. 748), which purports that the major symptoms of anorexia, namely food restriction, hyperactivity, and starvation denial, were adaptive in the EEA under conditions of famine. She thinks that the confluence of these symptoms may have been advantageous to those anorexic individuals to seek out new locations for food. Further, she reasons that under conditions of feast rather than famine, those who could store excess calories as fat would have been at an advantage over those who could not, if famine followed feast closely. This "thrifty genotype" is obviously harmful in the present environment, but eating is clearly the most essential of human motivations, and it remains with appetite, as Guisinger notes, a critical component, along, of a "highly regulated and tuned physiological and behavioral systems" (p. 748).

Evolutionary psychologist Laith Al-Shawaf (personal communication) has proposed that eating disorders may be partly caused by evolved mechanisms designed to afford success in mating competition because modern humans are exposed to hyperattractive models and celebrities, which are held up as the "ideal." He reasons that modern humans are not well designed to process such information because small group sizes in the EEA meant there were far fewer women available for mating, especially because many were already mated, pregnant, or breastfeeding (therefore, not ovulating). In contemporary societies, people are exposed to many more attractive strangers, both by images and in person. Al-Shawaf argues that interacting with such high numbers of attractive people may be totally unnatural and novel, given human evolutionary history. Further, modern media allows the manipulation of images, which increases the perception that the world is full of attractive, potential mates. Al-Shawaf reasons that our evolved mechanisms are not designed to manage this deluge of attractiveness, and he thinks it may be one of the causes of the intense intersexual mating competition and intrasexual mating competition, with the result being eating disorders.

Emotions

Throughout this chapter's discussion on the adaptive nature of psychopathologies is the implied notion that emotions are *the* critical outlet for the expression of the various types of psychopathology. The evolution of emotions has long been a topic of interest, particularly since 1872, when Darwin published his book, *The Expression of Emotions in Man and Animals*, where he reasoned that human emotions are derived from animal behavior, they have a biological basis, their purpose is social communication, and the mental states associated with emotions are linked to body movements. I have also emphasized throughout this book that modern human brains follow the evolutionary trajectory first established in primates; that is, our common ancestor with primates evolved to navigate more complex social groups through enhanced communication, including the expression and monitoring of emotional states.

More recently, evolutionary psychologists Al-Shawaf, David Buss, and their colleagues (Al-Shawaf, Conroy-Beam, Asao, & Buss, 2015; Lewis, Al-Shawaf, Conroy-Beam, Asao, & Buss, 2017) have reiterated and emphasized that human emotions evolved to solve a wider variety of adaptive problems than previously imagined. They hypothesize that the evolutionary function of emotions is to serve as *superordinate mechanisms*, which coordinate and regulate a host of functions, including basic perceptions and attention, learning, short- and long-term memory, motivations and their goals and effort allocation, inferences regarding categorizations and conceptual frameworks, social communication and language use, and other behaviors (see Cosmides & Tooby, 2000, for additional activities thought to be regulated by emotions). One of the unique contributions by Buss and his colleagues is that they address emotions that had been typically neglected by traditional evolutionary approaches. Traditional approaches focused on what might be considered *basic* emotions, like fear, anger, and disgust, while little attention had been given to love, either romantic, sexual, or parental. It is important to note that the very existence of a set of *basic* emotions is a subject of great debate. Overall, Buss et al. did emphasize that emotions appear to have evolved to solve the problems related to reproductive success, so I shall first focus my discussion on love.

What's Love Got to Do with It?

I previously examined the emotions of paranoia and jealousy with regard to the differing reproductive strategies of men and women. The evolutionary foundations of love are somewhat less clear than for jealousy and paranoia. Further, some evolutionary psychologists appear cynical, postulating that love evolved to blind people to true intentions. I once attended the fourth wedding of an acquaintance (and her third a few years earlier). She publicly repeated her oft-made promise to love and honor her spouse until death do they part. I quietly suggested to my wife that her vow should be modified to "love and honor until something better came along." But I digress: How can a man promise to be sexually faithful to one woman for his entire adult life when a majority of married men admit to being sexually unfaithful?

The standard evolutionary strategy from a man's perspective is getting his genes into the gene pool, which involves convincing a woman that he is trustworthy, sexually faithful, and will make a good, reliable father to their children. Evolutionary psychologists propose that women have evolved mechanisms to detect lying and ambivalence. It is in their evolutionary interest not to be fooled by an overnight flame or one who will be an absent or neglectful father. Women must be selective and gauge their sperm donor's MPI. A woman's life and her children's lives (or her genes' destiny) depend on the proper choice. She will be able to detect insincerity, fickleness, and lying. According to evolutionary psychologists, in order to break through these impressive defenses, a man will need a super-weapon: love.

Evolutionary biologist Robert Trivers (1972) argued that one effective way to deceive someone is to actually believe in what you are saying. Thus, if a man truly believed that he could make a lifetime commitment and he told the woman this, then a woman would be more likely to believe such a statement. In this scenario, love evolved as a form of self-deception, and it is used by men (mostly unconsciously) to circumvent a woman's defenses against deception. I hypothesize that a woman's love is more complicated, at least in terms of her love for her man. Indeed, if we tried to answer singer Tina Turner's original question, "What's love got to do with it?" we would have to conclude that, from a woman's perspective, the answer is "nothing." She needs a man's sperm. She needs his continued help and support in raising the children, but there is no ultimate need to "love" the man that provides these things. She has the precious eggs. He has the inexpensive sperm. At this point, love appears to have nothing to do with mating or reproduction.

However, a woman who "loves" her children will have an evolutionary fitness advantage over women with lesser maternal feelings. Thus, the origins of a woman's love may reside in her devotion to her children. Indeed, natural selection has seen to it that maternal devotion is strong in most mammals, whose young are more vulnerable at birth than those of non-mammals. In fact, maternal devotion appears to be correlated with the number of young. Evolutionary scientists have categorized maternal behavior into two strategies: r-selection, for species who produce large numbers of offspring, like turtles, whose dozens of babies are on their own after breaking out of their shells, and K-selection, for species that produce very few offspring and in which maternal devotion and care are paramount. Thus, it seems possible that a woman's love for her spouse might have its basis in maternal devotion or maternal love, and it is not hard to imagine the parallel between the way a woman expresses her love for her children and her love for her spouse. It also follows that if an ancestral couple shared feelings of love, compared to another couple where love was not equally shared, that shared love might be at an evolutionary fitness advantage. In the animal kingdom where there are adult attachments (mother and father bonding), there is a strong relationship between adult attachment and parental investment. Therefore, it appears that another function of love is to motivate parental interest and investment in the children.

Disgust: Bringing It All Back Home

I have discussed a variety of emotions and emotional states that accompany various forms of psychopathology which factor heavily in their adaptive nature. Al-Shawaf and his colleagues (2016) have stated that emotions are adaptations that evolved to solve the challenges of everyday life: obtaining nutritious food, avoiding bad food and infections, getting good mates, rearing their young, and avoiding predation for themselves and their young. They also note that these various adaptations require coordination, because some adaptations may conflict with others. They give the example of approach-avoidance motivations, wherein a search for a reproductive mate might invoke an approach motivation but an infected sore might invoke avoidance of that potential mate. They labeled this conflict of outputs the *coordination* problem, whereby random, haphazard, or unsystematized applications of concurrent adaptations would be much less likely to solve the many challenges of life. Cogently, they note: "The problem of conflicting outputs can be solved by mechanisms that control *activation* and *deactivation* of programs so as to minimize interference with effective problem solving" (italics mine; Al-Shawaf et al., 2016, p. 714).

I propose that their referent mechanisms are the cognitive functions that co-evolved with the brain structures discussed throughout this book. In this manner, emotions regulate cognitive processes (and vice versa), and the brain has been the ultimate organism for resolving these conflicting outputs. More importantly, it is interesting that they have chosen, as I highlighted, activation and deactivation as the principal means of program coordination. I have emphasized throughout this book that the essential principles guiding life forms rest on learning and memory systems, in particular non-associative learning, that is, sensitization and habituation. Thus, my interest in Al-Shawaf et al.'s principle means of emotion coordination is that it supports my hypothesis that *activation* is a form of *sensitization* and *deactivation* is a form of *habituation*.

Summary

1. Paleopsychopathology is the study of mental problems and mental diseases that may have increased relative fitness in the ancestral environment but do not enhance fitness in the present environment.
2. Some present psychopathologies (like Huntington's chorea) may not have had any adaptive value in the ancestral environment, as some genetic disorders' onset is well after what would have been prime reproductive years in the ancestral environment.
3. Some psychopathologies may not have been advantageous in and of themselves, but their polygenic basis may have given rise to successful adaptive phenotypes that are simply associated with those particular genes. For example, the location for the genes for schizophrenia are associated with the coding for immunity genes and creativity.
4. Modern personality disorders may have been adaptive in the ancestral environment because of their benefits in navigating social hierarchies. The evolution of an array of emotions may have also benefitted navigation in social hierarchies.

A Final Summary

Summarizing brain evolution is a Herculean task, and perhaps even a Sisyphean task, given the burgeoning literature on brains and their functions afforded by modern physiological methods and technology. Nevertheless, I think some summary statements are possible.

1. The evolution of life has been dependent on basic learning and memory systems. Even the evolution and nature of RNA and DNA has been the storage and recall of programs that solve life's challenges.
2. Learning and memory systems began with non-associative learning principles, sensitization and habituation. I hypothesize that these principles were ultimately guided by chemical affinities and disaffinities, which themselves are based on atomic and subatomic affinities and disaffinities.
3. Associative learning principles such as classical and operant conditioning appear to have led to an explosion in the diversity of living

organisms, which can be characterized as a "speciatic war" that marks the Cambrian explosion about 540 million years ago. If Bronfman, Ginsburg, and Jablonka (2016) are correct, then virtually unlimited associative learning at that time marked a transition to *conscious* life entities. I also propose that these associative learning principles represented exaptations of non-associative learning principles.

4. It also appears that the activation and deactivation programs of organisms to solve life's challenges mirrored the basic principles of non-associative learning; that is, activation is a form of sensitization, and deactivation is a form of habituation.

5. *Homo sapiens'* brains reached some zenith in size about 300,000 to 100,000 years ago. Since then, overall brain size has diminished about 10%, without any evidence of diminished cognitive function. In *Homo sapiens,* the parietal lobes and the cerebellum appear to have increased in relative size over this same time period. It is interesting to wonder whether these brain growth and diminishment trends will continue and what their consequences might be.

6. Finally, it is interesting to speculate whether the higher cognitive exaptations of the parietal lobes and the cerebellum played primary roles in modern symbolic thinking and modern language. Further, it appears that the olfactory bulbs have enlarged in modern *Homo sapiens* compared to those in our cousins the Neandertals. The olfactory bulbs have no established cognitive functions; however, their role in emotions like disgust and their critical role in modern immune systems are monumental. Olfaction may have also played an important part in mate selection and bonding. Will this olfactory bulb expansion trajectory continue as well, and what are those repercussions?

Coda

Historian Yuval Harari (2017) provocatively pondered the future of humanity. I have just laid out some possible brain trajectories for modern *Homo sapiens* that are based on the past 250,000 years: smaller overall brains; expanding parietal lobes, cerebellum, and olfactory bulbs; and diminishing occipital lobes. It is nearly impossible to predict whether these evolutionary trends will continue and what the consequences will be for cognition and emotion. Harari even noted it is difficult to understand the full array of mental states possible in modern *Homo sapiens*. As evidence for this limitation, he showed that a strong majority of psychological research (primarily social and personality psychology) is conducted on highly educated and relatively wealthy people in Western industrialized and democratic cultures. He also speculated that Neandertals, with larger brains than those of modern *Homo sapiens*, may have been capable of mental states that modern people cannot experience. He gave the example of philosopher Thomas Nagel's (1974) classic paper, "What Is It Like to Be a Bat?" In it, Nagel proposed that it is impossible for humans to know what it *feels* like to be bat. He wrote that we can study bats, analyze the echolation system of their movement and flight, categorize their behaviors, and explore them anatomically, but we cannot ever know what a bat actually feels. In this same vein, the philosopher Ludwig Wittgenstein (1953) elusively stated, "If a lion could speak, we could not understand him" (p. 223). As has been noted by many, the statement creates a paradox, because whatever language is used, it can be translated into another language. However, it is possible that Wittgenstein's main point may have been, as experienced by anyone who has attempted to translate a poem into another language, that there exists a basic incommensurability among languages. Further, there may exist an incommensurability among users of the same language, as in between young and old people, between rich and poor, and among people of different ethnicities and cultures. Indeed, there may reside a critical incommensurability between the lives and languages of Neandertals and *Homo sapiens*. This is not to say we were not alike, as we probably shared many more similarities than differences (Wynn, Overmann, & Coolidge, 2016). And yet, Neandertals became extinct. Was that event a mere twist of fate, or did our smaller brains and smaller occipital lobes, but expanded parietal lobes, larger temporal lobes, larger cerebellum, and larger olfactory bulbs, have cognitive and behavioral consequences that allowed us to survive and them to perish? What did the last Neandertals think as they watched these gracile and less robust *Homo sapiens* invade their territories? What did the last Neandertals ponder, perhaps sitting in a cave or on a rock overhang, as they looked out over the vast ocean? Of course, we cannot even fathom their thoughts. To ponder near and distant brain evolutions and their cognitive and emotional sequelae is truly a daunting task. Yet, nevertheless, it is intriguing, inviting, and inspires wonder.

Glossary

Action perception circuits areas in the brain where motor and sensory neurons converge and influence each others' functions. Also called *convergence zones*.

Adaptation physical or behavioral features which, through natural selection, aided survival and reproduction.

Alpha waves electrical activity of the brain with a frequency of 8–13 Hz that occurs in humans when conscious, relaxed and eyes are closed.

Amygdala subcortical brain region involved with the experiencing of emotions, particularly fear.

Analog comparison one of the two core processes in numerosity; it is the ability to differentiate between smaller and larger sets of things.

Angular gyrus brain region in the inferior parietal lobes that are critical in mathematical operations and many other higher cognitive functions.

Anterograde amnesia an inability to learn new information; critically dependent on the hippocampus.

Aphasia loss of ability to understand or express speech.

Apophenia the tendency for humans to attach meaning or perceive meaningful patterns to events that are random or not meaningful.

Apoptosis the genetically programmed death of brain cells that occurs as a normal part of an organism's growth or development.

Articulatory loop a part of the phonological component of Baddeley's working memory model; it can maintain and/or rehearse verbal or acoustic information either vocally or subvocally.

Associative learning one of the two major forms of learning, appearing about 545 million years ago, consisting of classical conditioning and operant conditioning. See *nonassociative learning*.

Astrocyte a star-shaped type of glial cell in the brain.

Australopithecus afarensis an extinct hominin appearing about 4 million years ago to about 3 million years ago; consisting of many species so they are collectively referred to as the australopithecines.

Autism spectrum disorder (ASD) a multisystem disorder that affects the brain, immune system, and gastrointestinal tract particularly affected are social and emotional functioning.

Autobiographical memory the recall of a personal event or scene with a strong sense of self in the memory, thus restricted to humans.

Autonoesis the human awareness that time is relative; one can imagine past, present, or the future in alternative or counterfactual scenarios.

Axon the long, threadlike part of a nerve cell along which impulses are conducted from the cell body to terminal synaptic vesicles to other cells.

Basal ganglia a group of structures in the base of the cerebrum, connected to the thalamus and cerebral cortex, that are involved in coordination of movement and higher cognitive functions.

Beta waves electrical activity of the brain with a frequency of 18–25 Hz and that can occur during wakefulness or sleep.

Broca's aphasia loss of speech fluency, trouble speaking, usually as a result of injury to frontal brain regions.

Broca's area a region of the inferior frontal lobes associated with the production of speech, temporal sequencing, and other cognitive processes; also activated during stone tool production.

Central executive a component of Baddeley's working memory model that acts as a supervisory and attentional monitor of behavior, including such functions as inhibition and decision-making; synonymous with the concept of executive functions of the frontal lobes.

Cerebellar cognitive affective syndrome a collection of symptoms associated with clinical or subclinical damage to the cerebellum what includes problems with executive functions, deficient abstract reasoning, decreased verbal fluency, impaired visuospatial cognition, inappropriate behaviors, and language difficulties.

Cerebellum a posterior and inferior cortical region whose function is to coordinate and regulate muscular activity and to control, regulate, and refine thinking.

Cerebrum refers to the whole brain, including the cortex, midbrain, and lower brain structures.

Circadian rhythm a day–night, 24-hour cycle that regulates many physiological processes.

Classical conditioning a type of associative learning in which a neutral stimulus is paired with an unconditioned stimulus (US) and eventually becomes a conditioned stimulus (CS) to elicit a conditioned response (CR).

Concerted evolution occurs where natural selection acts on functionally related and interconnected brain regions. See *mosaic evolution*.

Cortex (plural *cortices*) the upper part of the cerebrum, in mammals the upper six layers of cells; synonymous with the term *neocortex*.

Cytoarchitecture how cells, particularly in the central nervous system and the cerebrum, are arranged or layered.

Declarative memory the ability to learn and recall facts, meanings, and details; also called sematic memory and or explicit memory.

Default mode network brain activity when a human or nonhuman primate is resting and not engaged in a specific mental activity.

Delta waves high-amplitude brain waves with a frequency of 0.5 to 3 Hz and present in slow-wave sleep of most animals.

Dendrite the bushy-like extensions of a neuron's cell body; they take up chemical neurotransmitters in synapses and transmit those signals to the cell body.

Denisovans an extinct species of *Homo* appearing about 700,000 years ago.

Encephalization Quotient (EQ) a relative measure of brain size; it is the ratio of brain size to body size.

Epigenetics the study of the non–DNA-linked forms of inheritance, such as alterations of chromatin, the attachment of methyl groups to existing DNA, and gene–environment interactions.

Episodic buffer a component of Baddeley's working memory model; serves as a temporary memory store for the central executive and integrates information from the two subsystems, phonological storage and visuospatial sketchpad.

Episodic memory the memory of scenes and events with what, where, and when characteristics and sufficient emotional valence; attributed to humans and other animals.

Eukaryote a cell with a nucleus containing DNA and/or RNA.

Evolutionary neuropsychology the study of the evolution of the structures and functions of the brain.

Exaptation the change in phenotypic function of a pre-existing structure or feature from its original adaptive value to a new phenotype.

Executive functions a set of cognitive processes associated with the prefrontal cortices, which control, monitor, and inhibit behavior; processes include directing attention, attaining goals, decision-making, planning, organizing, and forming strategies.

Explicit memory the recall of facts, words, meanings, and events.

Extant still in existence as opposed to extinct.

Fasciculus (plural *fasciculi*) bundles of nerve fibers that transmit signals across brain regions.

Flashbulb memory a subtype of autobiographical memory, which only humans possess because there is a strong sense of self in the memory; the highly detailed and vivid recall of an event that was strongly emotionally arousing, such as a trauma, accidents, disasters, or a happy or wonderful event.

Frequency-dependent selection the value of a trait will decrease or increase as its frequency in the population increases or decreases.

Frontal lobes the largest lobes of the brain in the anterior region of the cerebrum, anterior to the central sulcus, and superior to the lateral fissure; associated with executive functions and the central executive in Baddeley's working memory model. Damage to the frontal lobes usually involves severe apathy, a loss of spontaneity, and impairment of the executive functions.

Gamma waves electrical activity of the brain with a frequency of 25–100 Hz, usually appearing in light sleep, with a low amplitude and anirregular pattern.

Genotype the entire set of genes of an organism.

Great apes a taxonomic family called Hominidae, which includes bonobos, chimpanzees, gorillas, and orangutans, *Homo sapiens,* and their extinct relatives such as *Homo erectus* and Neandertals.

Gyrus (plural *gyri*) the ridges of the cortex and lined by sulci, which are shallow indentations between ridges.

Habituation one of the two types of non-associative learning, in which the strength or probability of a response is reduced by repeated presentation of a stimulus; in essence, it means learning not to respond to irrelevant stimuli. See *sensitization.*

Hippocampus a brain region located bilaterally within the temporal lobes whose adaptation was the memorization of locations and smells and whose major exaptation is the formation of declarative memories.

Hominid any member of the family of *Hominidae*; includes all extant and extinct great apes, including humans and all their extinct relatives.

Hominin includes all extant and extinct humans and all of their immediate and distant ancestors, such as the australopithecines and the habilines, within about the last 6 million years.

Homo erectus an extinct species of *Homo,* larger brained than any previous hominin, with sophisticated stone tools (handaxes), appearing about 1.9 million years ago.

Homo habilis an archaic species of *Homo,* likely a cousin of the australopithecines but bigger brained, appearing about 2.5 million years ago and persisting until about 1.5 million years ago; the different species are collectively referred to as the habilines.

Homo heidelbergensis an archaic species of *Homo,* larger brained than any previous hominin, appearing about 800,000 years ago; the reputed ancestor of *Homo sapiens* and Neandertals.

Implicit memory the memory for skills or motoric procedures, such as juggling, bicycle riding, and stone knapping.

Intraparietal sulcus located in the parietal lobes and plays a critical role in numerosity; has specific single neurons and bundles of neurons that respond to symbolic and non-symbolic numbers.

K-complex a large and initially negative deflection in an electroencephalographic (EEG) waveform that occurs regularly during humans' stage 2 sleep.

K-selection A naturally selected evolutionary strategy where a species produces a few but "expensive" offspring. See *r-selection.*

Lobotomy a surgical procedure operation involving incisions into frontal lobes of the brain, which destroys connections within each frontal lobe; formerly used to treat disorders such as schizophrenia and violent psychiatric patients but now rarely conducted.

Long-term memory the recall of learned material over minutes, hours, days, or years; can refer to explicit/declarative memories or implicit/procedural memories.

Mentalizing a cognitive process of imagining; it is called theory of mind when a person can estimate the thoughts, attitudes, or emotions of other people.

Mirror neurons cells that fire both when an animal acts and when the animal observes the same action performed by another.

Mosaic evolution occurs where natural selection acts on individual brain regions. See *concerted evolution*.

Muscle atonia during human and nonhuman primate REM sleep, it is the paralysis of voluntary skeletal muscles so that dreams are not acted out physically.

Neandertals an extinct hominin, appearing about 400,000 years ago to about 30,000 years ago; largest brains of any hominin including modern *Homo sapiens*; subtle brain differences resulting in cognitive differences between Neandertals and *Homo sapiens* may have led to their extinction.

Negative frequency–dependent selection the value of a trait will decrease as its frequency in the population increases.

Neural reuse a hypothesis that brain neurons may be repurposed, recycled, or reused to serve purposes other than those for which they were originally adapted. See *exaptation*.

Neuron a specialized cell in the central nervous system (nerve cell) that transmits electrochemical impulses to other brain or central nervous system cells.

Neuroplasticity the limited ability of the brain to form and reorganize synaptic connections after learning or injury. There appears to be greater neuroplasticity before puberty than after.

Neurotransmitter chemicals in the synaptic vesicles of a neuron's dendrites, such as dopamine, norepinephrine, serotonin, and others, that when signaled to fire, excrete their contents into the synaptic gap between the dendrites and the adjacent axons. These axons have postsynaptic vesicles, which receive their signals only from specific neurotransmitters.

Non-associative learning one of the two major forms of learning, appearing about 3.9 billion years ago with the first forms of life; the two types of non-associative learning are habituation and sensitization. See *associative learning, habituation, sensitization*.

Numerosity the appreciation of numbers with two core processes: subitization (differentiating among one, two, and three things) and analog comparisons (discrimination between small and large sets). Numerosity may serve as the cognitive basis for abstractive thinking.

Occam's razor a simpler explanation is preferred over more complex ones, and an explanation with fewer assumptions is preferred over one with more assumptions.

Occipital lobes located in the posterior portion of the cerebrum and responsible for primary visual processing. In modern *Homo sapiens* they have diminished in size compared to that in Neandertals

Olfaction the sense of smell.

Olfactory bulbs bilateral structures inferior to the frontal lobes and responsible for the detection of odors. They are larger in modern *Homo sapiens* than in Neandertals.

Operant conditioning one type of associative learning, in which behavior is controlled by its consequences. Both positive reinforcement and negative reinforcement increase the probability of a response, whereas punishment decreases the likelihood of a response.

Over-determined an archaeological term for handaxes where the knappers spent more time than was needed to make a useful tool.

Overdetermination used to describe a phenomenon that has more than one cause.

Paleopsychopathology the study of mental problems and syndromes in the present environment, which may have evolved because they were adaptive in the ancestral environment.

Pareidolia a tendency for humans to perceive a face or pattern where one actually does not exist. See *apophenia*.

Parietal lobes regions of the cerebrum posterior to the frontal lobes and anterior to the occipital lobes; expanded in size in the recent evolution of *Homo sapiens'* brains compared to their size in Neandertals. They process sensory information in animals and humans, including sense of self in the latter, and they have a role in episodic memory in all animals.

Personality disorders cross-cultural, chronic, pervasive, and inflexible patterns of behavior under strong genetic influence. They usually begin in childhood or adolescence and lead to significant distress or impairment in occupational, educational, and social interactions.

Phenotype observable structural (physical) or behavioral characteristics of an organism resulting from the interaction of its genotype and environment.

Phonological loop (or store) a subsystem component of Baddeley's working memory model, which temporarily stores (about 2 seconds) verbal or acoustic information for additional processing. An articulatory processor can subsequently rehearse that information either vocally or subvocally for long-term memory storage.

Phrenology a discredited 1800s theory of Franz Gall, which proposed that the shape of the skull and skull bumps reflected a person's character and mental abilities.

Polygenic caused by two or more genes; often a large number of genes with each having a small effect.

Prefrontal cortex consists of the front part of the frontal lobes, critical in the executive functions.

Priming occurs where exposure to stimuli enhances or changes later performance, regardless of whether there was conscious recognition of the original stimuli.

Procedural memory an ability to learn and recall motor skills; a type of implicit memory.

Prokaryote the first single-celled organism without a distinct nucleus; appeared about 3.9 billion years ago.

Prosopagnosia an inability to recognize faces, usually due to damage to the fusiform area of the inferior temporal lobe.

Protolanguage an earlier and simpler form of modern languages, which may have initially consisted of warnings and expression of emotions (exclamatives), naming things (declaratives), and commands to others (imperatives).

Protozoan a single-celled eukaryote, considered the precursor of all later animals.

Recursion embedding a phrase within a phrase; necessary for subjunctive speech.

REM sleep also known as *rapid-eye-movement sleep;* in humans it occurs primarily during the last third of a typical sleep period; is associated with vivid dreaming and has been identified in all mammals, and some reptiles, fishes, and birds.

Retrograde amnesia an inability to recall previously learned information.

Retrosplenial cortex plays a central role in a network of brain regions for navigation, novel environments, and spatial memories.

RNA world hypothesis purports that the earliest form of life was a single strand of RNA that could self-metabolize and replicate.

r-selection A naturally selected evolutionary strategy where a species produces many but "inexpensive" offspring.

Sensitization one of the two types of non-associative learning, in which repeated stimuli result in increasingly stronger responses. It has been suggested that the definition of sensitization should also include steady or reliable responses to repeated stimuli.

Short-term memory the recall of learned material over short periods of time, such as seconds or minutes.

Sleep spindle brain waves consisting of short bursts of 13–16 Hz and that occur regularly during human's stage 2 sleep; hypothesized to play a role in learning and memory.

Slow-wave sleep (SWS) high-amplitude brain waves with a frequency between 0.5 and 3 Hz, also called *delta waves,* and are associated with stages 3 and 4 deep sleep. They have been observed in most animals, including many reptiles and all mammals; in humans, SWS is not associated with dreaming or thoughts.

Stone knapping the shaping of stones by other stones, bones, or wood to make sharp flakes, handaxes, choppers, or other stone tools. The earliest stone tools (Mode 1) were sharp flakes, appearing about 3.3 million years ago.

Subjunctive speech expresses hypothetical states, events or situations that may not have occurred, or ideas contrary to fact

Subitization an innate ability, in human infants and nonhuman primates, to differentiate quickly between one, two, and three things. There are neurons dedicated to subitization in the parietal lobes and medial temporal lobes.

Superordinate mechanism refers to the function of emotions, which coordinate and regulate a host of cognitive functions including language, attention, perception, and learning and memory.

Synapse a gap between the axon of a cell and the dendrites of an adjacent cell where information is passed from one neuron to another by chemical neurotransmitters.

Synaptic homeostatic hypothesis a process occurring during sleep where synapses renormalize and restore their ability to learn.

Synesthesia a process in which stimulation of one sense leads to the automatic co-stimulation of another sense, such as the perception of different numbers yielding the perception of particular colors, or musical notes eliciting colors.

Temporal lobes bilateral regions of the cerebrum, inferior to the frontal and parietal lobes; critical to sound processing in nonhuman primates and in speech and language in humans.

Theory of mind the ability to impute the thoughts and feelings of others.

Theta waves brain waves with a frequency of 4–6 Hz occurring in humans in stage 1 sleep; they are thought to be critical for spatial processing in animals.

Visuospatial sketchpad a component of Baddeley's working memory model responsible for visual and spatial perceptions of one's environment and the recall of visual and spatial memories.

Wernicke's aphasia the inability to understand language in its written or spoken form, usually due to damage to the superior and posterior temporal lobes.

Wernicke's area a brain region in the posterior portion of the superior gyri of the temporal lobes and thought to be responsible for comprehension of language.

Working memory (1) the ability to hold information in mind for processing in spite of interference; (2) Baddeley's multi-component model of cognition, which posits a central executive, episodic buffer, phonological storage, and a visuospatial sketchpad.

References

Aboitiz, F. (2017). *A brain for speech: A view from evolutionary neuroanatomy.* London, UK: Springer Nature.

Aboitiz, F., Aboitiz, S., & Garcia, R. R. (2010). The phonological loop. *Current Anthropology, 51*(S1), S55–S65.

Aboitiz, F., Garcia, R. R., Bosman, C., & Brunetti, E. (2006). Cortical memory mechanisms and language origins. *Brain and Language, 98*(1), 40–56.

Addis, D. R., Wong, A. T., & Schacter, D. L. (2007). Remembering the past and imagining the future: Common and distinct neural substrates during event construction and elaboration. *Neuropsychologia, 45*(7), 1363–1377.

Agustí, J., & Rubio-Campillo, X. (2016). Were Neanderthals responsible for their own extinction? *Quaternary International, 431*, 232–237.

Allen, T. A., & Fortin, N. J. (2013). The evolution of episodic memory. *Proceedings of the National Academy of Sciences of the United States of America, 110*(S2), 10379–10386.

Al-Shawaf, L., Conroy-Beam, D., Asao, K., & Buss, D. M. (2015). Human emotions: An evolutionary psychological perspective. *Emotion Review, 8*(2), 173–186.

Al-Shawaf, L., Zreik, K., & Buss, D. M. (2018). Thirteen misunderstandings about natural selection. *Encyclopedia of Evolutionary Psychological Science*, 1–14.

Altarelli, I., Leroy, F., Monzalvo, K., Fluss, J., Billard, C., Dehaene-Lambertz, G., . . . Ramus, F. (2014). Planum temporale asymmetry in developmental dyslexia: Revisiting an old question. *Human Brain Mapping, 35*(12), 5717–5735.

American Psychiatric Association. (2013). *Diagnostic and statistical manual of mental disorders* (5th ed.). Washington, DC: Author.

Anderson, M. L. (2010). Neural reuse: A fundamental organizational principle of the brain. *Behavioral and Brain Sciences, 33*(4), 245–266.

Anderson, M. L. (2014). *After phrenology: Neural reuse and the interactive brain.* Cambridge, MA: MIT Press.

Aron, A. R., Robbins, T. W., & Poldrack, R. A. (2004). Inhibition and the right inferior frontal cortex. *Trends in Cognitive Sciences, 8*(4), 170–177.

Arsuaga, J. L. (2002). *The Neanderthal's necklace: In search of the first thinkers.* New York, NY: Four Walls Eight Windows.

Aserinsky, E., & Kleitman, N. (1953). Regularly occurring periods of eye motility, and concomitant phenomena, during sleep. *Science, 118*(3062), 273–274.

Attout, L., Fias, W., Salmon, E., & Majerus, S. (2014). Common neural substrates for ordinal representation in short-term memory, numerical and alphabetical cognition. *PloS One, 9*(3), e92049.

Auger, S. D., Zeidman, P., & Maguire, E. A. (2015). A central role for the retrosplenial cortex in de novo environmental learning. *Elife, 4*, e09031.

Augustinack, J. C., van der Kouwe, A. J., Salat, D. H., Benner, T., Stevens, A. A., Annese, J., . . . Corkin, S. (2014). HM's contributions to neuroscience: A review and autopsy studies. *Hippocampus, 24*(11), 1267–1286.

Baddeley, A. D. (2000). The episodic buffer: A new component of working memory? *Trends in Cognitive Sciences, 4*(11), 417–423.

Baddeley, A. D. (2001). Is working memory still working? *American Psychologist, 56*(11), 851–864.

Baddeley, A. D. (2012). Working memory: Theories, models, and controversies. *Annual Review of Psychology, 63*, 1–29.

Baddeley, A., Gathercole, S., & Papagno, C. (1998). The phonological loop as a language learning device. *Psychological Review, 105*(1), 158–173.

Baddeley, A. D., & Hitch, G. J. (1974). Working memory. In G. A. Bower (Ed.), *Recent advances in learning and motivation* (pp. 47–90). New York, NY: Academic Press.

Baddeley, A., & Logie. R. (1999). Working memory: The multi-component model. In A. Miyake & P. Shah (Eds.), *Models of working memory: Mechanisms of active maintenance and executive control* (pp. 28–61). New York, NY: Cambridge University Press.

Baldwin, J. M. (1896). A new factor in evolution. *The American Naturalist, 30*(354), 441–451.

Balsters, J. H., Whelan, C. D., Robertson, I. H., & Ramnani, N. (2013). Cerebellum and cognition: Evidence for the encoding of higher order rules. *Cerebral Cortex, 23*(6), 1433–1443.

Balter, M. (2017). Schizophrenia's unyielding mysteries. *Scientific American, 316*(5), 54–61.

Barton, R. A., & Venditti, C. (2014). Rapid evolution of the cerebellum in humans and other great apes. *Current Biology, 24*(20), 2440–2444.

Bastian, H. C. (1869). On the various forms of loss of speech in cerebral disease. *British and Foreign Medical and Chirurgical Review, 43*, 209–236.

Bastir, M., Rosas, A., Gunz, P., Peña-Melian, A., Manzi, G., Harvati, K., . . . Hublin, J. J. (2011). Evolution of the base of the brain in highly encephalized human species. *Nature Communications, 2*, 588.

Bechara, A., & Damasio, A. R. (2005). The somatic marker hypothesis: A neural theory of economic decision. *Games and Economic Behavior, 52*(2), 336–372.

Benchley, R. (1920, February). The most popular book of the month: An extremely literary review of the latest edition of the New York City telephone directory. *Vanity Fair*, 69.

Bennet, C., Baird, A., Miller, M., & Wolford, G. (2010). Neural correlates of interspecies perspective taking in the post-mortem Atlantic salmon: An argument for proper multiple comparisons correction. *Journal of Serendipitous and Unexpected Results, 1*(1), 1–5.

Bernstein, I. S., & Gordon, T. P. (1974). The function of aggression in primate societies: Uncontrolled aggression may threaten human survival, but aggression may be vital to the establishment and regulation of primate societies and sociality. *American Scientist, 62*(3), 304–311.

Bickerton, D. (2007). Language evolution: A brief guide for linguists. *Lingua, 117*(3), 510–526.

Bolhuis, J. J., Tattersall, I., Chomsky, N., & Berwick, R. C. (2014). How could language have evolved? *PLoS Biology, 12*(8), e1001934.

Bolnick, D. I., & Stutz, W. E. (2017). Frequency dependence limits divergent evolution by favouring rare immigrants over residents. *Nature, 546*(7657), 285–288.

Botha, R. P. (2002). Are there features of language that arose like birds' feathers? *Language & Communication, 22*(1), 17–35.

Botha, R. (2006). On the Windows Approach to language evolution. *Language & Communication, 26*(2), 129–143.

Botha, R. (2012). Protolanguage and the "God particle". *Lingua, 122*(12), 1308–1324.

Brain, C. K. (1972). An attempt to reconstruct the behaviour of australopithecines: The evidence for interpersonal violence. *African Zoology, 7*(1), 379–401.

Bramble, D. M., & Lieberman, D. E. (2004). Endurance running and the evolution of *Homo. Nature, 432*, 345–352.

Brawn, T. P., Fenn, K. M., Nusbaum, H. C., & Margoliash, D. (2010). Consolidating the effects of waking and sleep on motor-sequence learning. *Journal of Neuroscience, 30*(42), 13977–13982.

Breasted, J. (1991). *The Edwin Smith surgical papyrus: Hieroglyphic transliteration translation and commentary* (Vol. 1). Chicago, IL: University of Chicago Press. (Original work published 1930).

Bronfman, Z. Z., Ginsburg, S., & Jablonka, E. (2016). The evolutionary origins of consciousness: Suggesting a transition marker. *Journal of Consciousness Studies, 9-10*, 7–34.

Bruner, E. (2004). Geometric morphometrics and paleoneurology: Brain shape evolution in the genus *Homo. Journal of Human Evolution, 47*(5), 279–303.

Bruner, E. (2010). Morphological differences in the parietal lobes within the human genus. *Current Anthropology, 51*(S1), S77–S88.

Bruner, E., Amano, H., Pereira-Pedro, A. S., & Ogihara, N. (2018). The evolution of the parietal lobes in the genus *Homo.* In E. Bruner, N. Ogihara, & H. C. Tanabe (Eds.), *Digital endocasts: From skulls to brains* (pp. 219–237). Tokyo: Springer.

Bruner, E., & Holloway, R. L. (2010). A bivariate approach to the widening of the frontal lobes in the genus *Homo. Journal of Human Evolution, 58*(2), 138–146.

Bruner, E., & Iriki, A. (2016). Extending mind, visuospatial integration, and the evolution of the parietal lobes in the human genus. *Quaternary International, 405*, 98–110.

Bruner, E., & Lozano, M. (2014). Extended mind and visuo-spatial integration: Three hands for the Neandertal lineage. *Journal of Anthropological Sciences, 92*, 273–280.

Buckner, R. L., & Krienen, F. M. (2013). The evolution of distributed association networks in the human brain. *Trends in Cognitive Sciences, 17*(12), 648–665.

Burgess, N., Becker, S., King, J. A., & O'Keefe, J. (2001). Memory for events and their spatial context: Models and experiments. *Philosophical Transactions of the Royal Society of London B: Biological Sciences, 356*(1413), 1493–1503.

Burgess, N., Maguire, E. A., & O'Keefe, J. (2002). The human hippocampus and spatial and episodic memory. *Neuron, 35*(4), 625–641.

Cabestrero-Rincón, M. A., Balzeau, A., & Lorenzo, C. (2018). Differential evolution of cerebral and cerebellar fossae in recent *Homo:* A new methodological approach. *Journal of Comparative Human Biology, 69*, 289–303.

Cachel, S., & Harris, J. (1995). Ranging patterns, land-use and subsistence in *Homo erectus* from the perspective of evolutionary biology. In J. Bower and S. Sartono (Eds.), *Evolution and ecology of Homo erectus* (pp. 51–66). Leiden: Pithecanthropus Centennial Foundation.

Cai, D. J., Mednick, S. A., Harrison, E. M., Kanady, J. C., & Mednick, S. C. (2009). REM, not incubation, improves creativity by priming associative networks. *Proceedings of the National Academy of Sciences of the United States of America, 106*(25) 10130–10134.

Caligiore, D., Pezzulo, G., Baldassarre, G., Bostan, A. C., Strick, P. L., Doya, K., . . . Lago-Rodriguez, A. (2017). Consensus paper: Towards a systems-level view of cerebellar function: The interplay between cerebellum, basal ganglia, and cortex. *Cerebellum, 16*(1), 203–229.

Caplan, D., & Waters, G. S. (1995). On the nature of the phonological output planning processes involved in verbal rehearsal: Evidence from aphasia. *Brain and Language, 48*(2), 191–220.

Cavanna, A. E., & Trimble, M. R. (2006). The precuneus: A review of its functional anatomy and behavioural correlates. *Brain, 129*(3), 564–583.

Chance, S. A., Sawyer, E. K., Clover, L. M., Wicinski, B., Hof, P. R., & Crow, T. J. (2013). Hemispheric asymmetry in the fusiform gyrus distinguishes Homo sapiens from chimpanzees. *Brain Structure and Function, 218*(6), 1391–1405.

Cirelli, C., & Bushey, D. (2008). Sleep and wakefulness in *Drosophila melanogaster*. In D. W. Pfaff and B. L. Kieffer (Eds.), *Molecular and biophysical mechanisms of arousal, alertness, and attention* (pp. 323–329). Malden, MA: Blackwell Publishing.

Colón-Ramos, D. A. (2016). The need to connect: on the cell biology of synapses, behaviors, and networks in science. *Molecular Biology of the Cell, 27*(21), 3203–3207.

Comer, R. J. (2015). *Abnormal psychology* (9th ed.). New York, NY: Worth Publishers.

Coolidge, F. L. (2006). *Dream interpretation as a psychotherapeutic technique.* London, UK: Radcliffe.

Coolidge, F. L. (2012). On the emergence of grammatical language as a means of bypassing the limitations of working memory capacity. *Physics of Life Reviews, 9*, 217–218.

Coolidge, F. L. (2014). The exaptation of the parietal lobes in *Homo sapiens. Journal of Anthropological Sciences, 92*, 295–298.

Coolidge, F. L. (2018, September). *Evolutionary implications of the sense of numbers.* Poster presented at the meeting of the European Society for the study of Human Evolution, Faro, Portugal.

Coolidge, F. L., Estey, A., Segal, D., & Marle, P. D. (2013). Are alexithymia and schizoid personality disorder synonymous diagnoses? *Comprehensive Psychiatry, 54*, 141–148.

Coolidge, F. L., Merwin, M. M., Wooley, M. J., & Hyman, J. N. (1990). Some problems with the diagnostic criteria of the antisocial personality disorder in DSM-III-R: A preliminary study. *Journal of Personality Disorders, 4*(4), 407–413.

Coolidge, F. L., Middleton, P. A., Griego, J. A., & Schmidt, M. M. (1996). The effects of interference on verbal learning in multiple sclerosis. *Archives of Clinical Neuropsychology, 11*, 605–611.

Coolidge, F. L. & Overmann, K. A. (2012). Numerosity, abstraction, and the emergence of symbolic thinking. *Current Anthropology, 53*(2), 204–225.

Coolidge, F. L., Overmann, K. A., & Wynn, T. (2010). Recursion: What is it, who has it, and how did it evolve? *Wiley Interdisciplinary Reviews: Cognitive Science, 2*(5), 547–554.

Coolidge, F. L., & Segal, D. L. (1998). Evolution of personality disorder diagnosis in the *Diagnostic and Statistical Manual of Mental Disorders. Clinical Psychology Review, 18*(5), 585–599.

Coolidge, F. L., Segal, D. L., Klebe, K. J., Cahill, B. S., & Whitcomb, J. M. (2009). Psychometric properties of the Coolidge Correctional Inventory in a sample of 3,962 prison inmates. *Behavioral Sciences & the Law, 27*(5), 713–726.

Coolidge, F. L., Thede, L. L., & Jang, K. L. (2001). Heritability of personality disorders in childhood: A preliminary investigation. *Journal of Personality Disorders, 15*(1), 33–40.

Coolidge, F. L., & Wynn, T. (2001). Executive functions of the frontal lobes and the evolutionary ascendancy of *Homo sapiens. Cambridge Archaeological Journal, 11*(2), 255–260.

Coolidge, F. L., & Wynn, T. (2005). Working memory, its executive functions, and the emergence of modern thinking. *Cambridge Archaeological Journal, 15*(1), 5–26.

Coolidge, F. L., & Wynn, T. (2009). Recursion, phonological storage capacity, and the evolution of modern speech. In R. Botha & C. Knight (Eds.), *The prehistory of language* (pp. 244–254). Oxford, UK: Oxford University Press.

Coolidge, F. L., & Wynn, T. (2016). An introduction to cognitive archaeology. *Current Directions in Psychological Science, 25*(6), 386–392.

Coolidge, F. L., & Wynn, T. (2018). *The rise of* Homo sapiens (2nd ed.). New York, NY: Oxford University Press.

Coolidge, F. L., & Wynn, T. (2019, March). *The second cognitive Rubicon in stone-knapping: Late Acheulean handaxes.* Paper presented at the conference "Retuning cognition with a pair of rocks: Culture, evolution, technology," Center for Philosophy of Science, Cathedral of Learning, University of Pittsburgh, Pittsburg, PA.

Corballis, M. (2003). From hand to mouth: The gestural origins of language. In M. Christiansen & S. Kirby (Eds.), *Language evolution: The states of the art.* Oxford: Oxford University Press.

Corkin, S. (2013). *Permanent present tense: The unforgettable life of the amnesic patient, H.M.* New York, NY: Basic Books.

Cosmides, L., & Tooby, J. (2000). Evolutionary psychology and the emotions. *Handbook of Emotions, 2,* 91–115.

Crespi, B., Summers, K., & Dorus, S. (2007). Adaptive evolution of genes underlying schizophrenia. *Proceedings of the Royal Society of London B: Biological Sciences, 274*(1627), 2801–2810.

Crow, T. J. (1997). Is schizophrenia the price that *Homo sapiens* pays for language? *Schizophrenia Research, 28*(2), 127–141.

Crow, T. J. (2000). Schizophrenia as the price that *Homo sapiens* pays for language: A resolution of the central paradox in the origin of the species. *Brain Research Reviews, 31*(2), 118–129.

Crowder, R. G., & Morton, J. (1969). Precategorical acoustic storage (PAS). *Attention, Perception, & Psychophysics, 5*(6), 365–373.

Damasio, A. R. (1989). The brain binds entities and events by multiregional activation from convergence zones. *Neural Computation, 1*(1), 123–132.

Danjo, T., Toyoizumi, T., & Fujisawa, S. (2018). Spatial representations of self and other in the hippocampus. *Science, 359*(6372), 213–218.

Dart, R. A. (1949). The predatory implemental technique of *Australopithecus. American Journal of Physical Anthropology, 7*(1), 1–38.

Darwin, C. (1862). *On the various contrivances by which British and foreign orchids are fertilised by insects: And on the good effect of intercrossing.* Cambridge, UK: Cambridge Library Collection.

Darwin, C. (1872). *The expression of the emotions in man and animals.* London, UK: John Murray.

Deacon, T. W. (1996). Prefrontal cortex and symbol learning: Why a brain capable of language evolved only once. In B. M. Velichkovsky and D. M. Rumbaugh (Eds.), *Communicating meaning: The evolution and development of language* (pp. 103–138). Hillsdale, NJ: Lawrence Erlbaum Associates.

Dehaene, S. (2005). Evolution of human cortical circuits for reading and arithmetic: The "neuronal recycling" hypothesis. In S. Dehaene, J. R. Duhamel, M. Hauser, & G. Rizzolatti (Eds.), *From monkey brain to human brain* (pp. 133–157). Cambridge, MA: MIT Press.

Dehaene, S., & Cohen, L. (2007). Cultural recycling of cortical maps. *Neuron, 56*(2), 384–398.

Dehaene, S., Cohen, L., Sigman, M., & Vinckier, F. (2005). The neural code for written words: A proposal. *Trends in Cognitive Sciences, 9*(7), 335–341.

Delgado, M. R., Nearing, K. I., LeDoux, J. E., & Phelps, E. A. (2008). Neural circuitry underlying the regulation of conditioned fear and its relation to extinction. *Neuron, 59*(5), 829–838.

Dellinger-Ness, L. A., & Handler, L. (2006). Self-injurious behavior in human and non-human primates. *Clinical Psychology Review, 26*(5), 503–514.

de Lorenzo, V. (2014). From the selfish gene to selfish metabolism: Revisiting the central dogma. *BioEssays, 36*(3), 226–235.

DeLouize, A. M. (2017). *The contribution of running behavior to brain growth and cognition in primates.* (Unpublished master's thesis). University of Colorado, Colorado Springs.

DeLouize, A. M., Coolidge, F. L., & Wynn, T. (2017). Dopaminergic systems expansion and the advent of *Homo erectus. Quaternary International, 427*(Part B), 245–252. doi. org/10.1016/jquaint.2015.10.123

de Lussanet, M. H., & Osse, J. W. (2012). An ancestral axial twist explains the contralateral forebrain and the optic chiasm in vertebrates. *Animal Biology, 62*(2), 193–216.

de Lussanet, M. H., & Osse, J. W. (2015). Decussation as an axial twist: A comment on Kinsbourne (2013). *Neuropsychology, 29*(5), 713–714.

Deschamps, I., Baum, S. R., & Gracco, V. L. (2014). On the role of the supramarginal gyrus in phonological processing and verbal working memory: Evidence from rTMS studies. *Neuropsychologia, 53*, 39–46.

Dessalles, J. L. (2010). Have you anything unexpected to say? The human propensity to communicate surprise and its role in the emergence of language. In A. D. M. Smith, M. Schouwstra, B. de Boer, & K. Smith (Eds.), *The evolution of language—Proceedings of the 8th International Conference* (pp. 99–106). Singapore: World Scientific Publishing Co.

DeVore, I., & Tooby, J. (1987). The reconstruction of hominid behavioral evolution through strategic modeling. In W. G. Kinzey (Ed.), *The evolution of human behavior: Primate models* (pp. 183–237). Albany, NY: State University of New York Press.

Diekelmann, S., & Born, J. (2010). The memory function of sleep. *Nature Reviews Neuroscience, 11*(2), 114–126.

Dieringer, T., & Coolidge, F. L. (2018, September). *A review of recent Neandertal extinction theories.* Poster presented at the meeting of the European Society for the study of Human Evolution, Faro, Portugal.

Dittrich, L. (2017). *Patient H.M.: A story of memory, madness, and family secrets.* New York, NY: Random House.

Dodgson, G., & Gordon, S. (2009). Avoiding false negatives: Are some auditory hallucinations an evolved design flaw? *Behavioural and Cognitive Psychotherapy, 37*(3), 325–334.

Doya, K. (2000). Complementary roles of basal ganglia and cerebellum in learning and motor control. *Current Opinion in Neurobiology, 10*(6), 732–739.

Dubrovsky, B. (2002). Evolutionary psychiatry. Adaptationist and nonadaptationist conceptualizations. *Progress in Neuro-Psychopharmacology and Biological Psychiatry, 26*(1), 1–19.

Dunbar, R. I. (1998). The social brain hypothesis. *Brain, 9*(10), 178–190.

Dunbar, R. I. (2004). Gossip in evolutionary perspective. *Review of General Psychology,* *8*(2), 100–110.

Dunbar, R. I. (2013). *Primate social systems.* Berlin, Germany: Springer Science & Business Media.

Dunbar, R. I., & Shultz, S. (2007). Evolution in the social brain. *Science, 317*(5843), 1344–1347.

Durante, K. M., Griskevicius, V., Simpson, J. A., Cantu, S. M., & Li, N. P. (2012). Ovulation leads women to perceive sexy cads as good dads. *Journal of Personality and Social Psychology, 103*(2), 292–305.

Eisenstein, E. M., & Eisenstein, D. (2006). A behavioral homeostasis theory of habituation and sensitization: II. Further developments and predictions. *Reviews in the Neurosciences, 17*(5), 533–558.

Eisenstein, E. M., Eisenstein, D. L., & Sarma, J. S. M. (2016). An exploration of how to define and measure the evolution of behavior, learning, memory and mind across the full phylogenetic tree of life. *Communicative & Integrative Biology, 9*(3), e1166320.

Eklund, A., Nichols, T. E., & Knutsson, H. (2016). Cluster failure: Why fMRI inferences for spatial extent have inflated false-positive rates. *Proceedings of the National Academy of Sciences of the United States of America, 113*(28), 7900–7905.

Engle, R. W., & Kane, M. J. (2003). Executive attention, working memory capacity, and a two-factor theory of cognitive control. *Psychology of Learning and Motivation, 44,* 145–199.

Evans, P. D., Gilbert, S. L., Mekel-Bobrov, N., Vallender, E. J., Anderson, J. R., Vaez-Azizi, L. M., . . . Lahn, B. T. (2005). Microcephalin, a gene regulating brain size, continues to evolve adaptively in humans. *Science, 309*(5741), 1717–1720.

Evans, N., & Levinson, S. (2009). The myth of language universals: Language diversity and its importance for cognitive science. *Behavioral and Brain Sciences, 32*(5), 429–448.

Ferrer-i-Cancho, R. (2015). The placement of the head that minimizes online memory. *Language Dynamics and Change, 5*(1), 114–137.

Fiala, J. A., & Coolidge, F. L. (2018). Genetic predilections and predispositions for the development of shamanism. *Behavioral and Brain Sciences, 41,* e73.

Finger, S. (1994). *Origins of neuroscience: A history of explorations into brain function.* New York, NY: Oxford University Press.

Fitch, W., & Braccini, S. N. (2013). Primate laterality and the biology and evolution of human handedness: A review and synthesis. *Annals of the New York Academy of Sciences, 1288*(1), 70–85.

Fitch, W. T., Hauser, M. D., & Chomsky, N. (2005). The evolution of the language faculty: Clarifications and implications. *Cognition, 97*(2), 179–210.

Fletcher, P. C., Frith, C. D., Baker, S. C., Shallice, T., Frackowiak, R. S. J., & Dolan, R. J. (1995). The mind's eye—precuneus activation in memory-related imagery. *NeuroImage, 2*(3), 195–200.

Franklin, M. S., & Zyphur, M. J. (2005). The role of dreams in the evolution of the human mind. *Evolutionary Psychology, 3*(1), 59–78.

Freud, S. (1956). *The interpretation of dreams* (J. Strachey, Trans., Ed.). New York, NY: Basic Books. (Original work published 1900)

Freud, S. (1960). *Jokes and their relation to the unconscious* (J. Strachey, Trans., Ed.). London: W. W. Norton & Company. (Original work published 1905)

Friedman, N. P., Miyake, A., Young, S. E., DeFries, J. C., Corley, R. P., & Hewitt, J. K. (2008). Individual differences in executive functions are almost entirely genetic in origin. *Journal of Experimental Psychology: General, 137*(2), 201–225.

Fritsch, G., & Hitzig, E. (1870). Über die elektrische Erregbarkeit des Grosshirns. *Archiv für Anatomie, Physiologie und Wissenschaftliche Medicin, 37,* 300–332.

Gagliano, M., Abramson, C. I., & Depczynski, M. (2018). Plants learn and remember: Lets get used to it. *Oecologia, 186*(1), 29–31.

Gall, F. J., & Spurzheim, J. C. (1810). *Anatomie et physiologie* (4 vols.) Paris: Schoell, 1819.

Gannon, P. J., Holloway, R. L., Broadfield, D. C., & Braun, A. R. (1998). Asymmetry of chimpanzee planum temporale: Humanlike pattern of Wernicke's brain language area homolog. *Science, 279*(5348), 220–222.

Gánti, T. (1975). Organization of chemical reactions into dividing and metabolizing units: The chemotons. *BioSystems, 7*(1), 15–21.

Gärdenfors, P. (2013). The role of cooperation in the evolution of protolanguage and language. In G. Hatfield & H. Pittman (Eds.), *Evolution of mind, brain, and culture* (pp. 193–216). Philadelphia, PA: University of Pennsylvania Press.

Genzel, L., & Battaglia, F. P. (2017). Cortico-hippocampal circuits for memory consolidation: The role of the prefrontal cortex. In N. Axmacher & B. Rasch (Eds.), *Cognitive neuroscience of memory consolidation* (pp. 265–281). Cham, Switzerland: Springer International Publishing.

Ginsburg, S., & Jablonka, E. (2010). The evolution of associative learning: A factor in the Cambrian explosion. *Journal of Theoretical Biology, 266*(1), 11–20.

Ginsburg, S., & Jablonka, E. (2015). The teleological transitions in evolution: A Gántian view. *Journal of Theoretical Biology, 381,* 55–60.

Goldberg, E. (2002). *The executive brain: Frontal lobes and the civilized mind.* London, UK: Oxford University Press.

Goldberg, E. (2009). *The new executive brain: Frontal lobes in a complex world.* London, UK: Oxford University Press.

Goldman-Rakic, P. S., & Leung, H-C. (2002). Functional architecture of the dorsolateral prefrontal cortex in monkeys and humans. In D. T. Stuss & R. T. Knight (Eds.), *Principles of frontal lobe function* (pp. 85–95). London, UK: Oxford University Press.

Gordon, E. M., Lynch, C. J., Gratton, C., Laumann, T. O., Gilmore, A. W., Greene, D. J., . . . Dosenbach, N. U. (2018). Three distinct sets of connector hubs integrate human brain function. *Cell Reports, 24*(7), 1687–1695.

Gould, S. J., & Vrba, E. S. (1982). Exaptation: A missing term in the science of form. *Paleobiology, 8*(01), 4–15.

Green, R. E., Krause, J., Briggs, A. W., Maricic, T., Stenzel, U., Kircher, M., . . . Hansen, N. F. (2010). A draft sequence of the Neandertal genome. *Science, 328*(5979), 710–722.

Gross, C. G. (2008). Single neuron studies of inferior temporal cortex. *Neuropsychologia, 46*(3), 841–852.

Guisinger, S. (2003). Adapted to flee famine: Adding an evolutionary perspective on anorexia nervosa. *Psychological Review, 110*(4), 745–761.

Gunz, P., Neubauer, S., Golovanova, L., Doronichev, V., Maureille, B., & Hublin, J. J. (2012). A uniquely modern human pattern of endocranial development. Insights from a new cranial reconstruction of the Neandertal newborn from Mezmaiskaya. *Journal of Human Evolution, 62*(2), 300–313.

Harari, Y. N. (2017). *Homo deus.* New York, NY: HarperCollins.

Hargreaves, E. L., Rao, G., Lee, I., & Knierim, J. J. (2005). Major Dissociation Between Medial and Lateral Entorhinal Input to Dorsal Hippocampus. *Science, 308*(5729), 1792–1794.

Harlow, J. M. (1848). Passage of an iron rod through the head. *Boston Medical and Surgical Journal, 39,* 389–393.

Harlow, J. M. (1868). Recovery from the passage of an iron bar through the head. *Publications of the Massachusetts Medical Society, 2*(3), 327–346.

Harris, K., & Nielsen, R. (2016). The genetic cost of Neanderthal introgression. *Genetics, 203*(2), 881–891.

Hartmann, E. (1998). *Dreams and nightmares: The new theory on the origin and meaning of dreams.* New York, NY: Plenum Trade.

Harvey, B. M., Klein, B. P., Petridou, N., & Dumoulin, S. O. (2013). Topographic representation of numerosity in the human parietal cortex. *Science, 341*(6150), 1123–1126.

Haselton, M. G., & Gildersleeve, K. (2011). Can men detect ovulation? *Current Directions in Psychological Science, 20*(2), 87–92.

Hatemi, P. K., & McDermott, R. (2012). The genetics of politics: discovery, challenges, and progress. *Trends in Genetics, 28*(10), 525–533.

Hauser, M. D., Chomsky, N., & Fitch, W. T. (2002). The faculty of language: What is it, who has it, and how did it evolve? *Science, 298*(5598), 1569–1579.

Hazy, T. E., Frank, M. J., & O'Reilly, R. C. (2006). Banishing the homunculus: Making working memory work. *Neuroscience, 139*(1), 105–118.

Hedinger, R. L. (2016). *Epigenetics review and synthesis: Autism spectrum disorder, postpartum depression, and posttraumatic stress disorder* (undergraduate honors thesis). Ball State University, Muncie, IN.

Herculano-Houzel, S. (2012). The remarkable, yet not extraordinary, human brain as a scaled-up primate brain and its associated cost. *Proceedings of the National Academy of Sciences of the United States of America, 109*(Suppl. 1), 10661–10668.

Herculano-Houzel, S. (2015). Decreasing sleep requirement with increasing numbers of neurons as a driver of bigger brains and bodies in mammalian evolution. *Proceedings of the Royal Society B: Biological Sciences, 282*(1816), 20151853.

Herculano-Houzel, S. (2016). *Human advantage: A new understanding of how our brain became remarkable.* Cambridge, MA: MIT Press.

Herculano-Houzel, S., Manger, P. R., & Kaas, J. H. (2014). Brain scaling in mammalian evolution as a consequence of concerted and mosaic changes in numbers of neurons and average neuronal cell size. *Frontiers in Neuroanatomy, 8*(77), 1–28.

Hertkorn, N., Harir, M., Gonsior, M., Koch, B., Michalk, B., & Schmitt-Kopplin, P. (2013). Elucidating the biogeochemical memory of the oceans by means of high-resolution organic structural spectroscopy. In J. Xu, J. Wu, & Y. He (Eds.), *Functions of natural organic matter in changing environment* (pp. 13–17). Hangzhou, China: Springer-Zhejiang University Press.

Hickok, G. (2009). Eight problems for the mirror neuron theory of action understanding in monkeys and humans. *Journal of Cognitive Neuroscience, 21*(7), 1229–1243.

Hickok, G. (2014). *The myth of mirror neurons: The real neuroscience of communication and cognition.* New York, NY: W. W. Norton & Company.

Holloway, R. L. (1983). Human brain evolution: A search for units, models and synthesis. *Canadian Journal of Anthropology, 3*(2), 215–229.

Holmes, G. (1917). The symptoms of acute cerebellar injuries due to gunshot injuries. *Brain, 40*(4), 461–535.

Holmes, G. (1922). The Croonian lectures on clinical symptoms of cerebellar disease and their interpretation. *The Lancet, 200,* 59–65.

Holmes, G. (1939). The cerebellum of man. *Brain, 62*(1), 1–30.

Hoppenbrouwers, S. S., De Jesus, D. R., Stirpe, T., Fitzgerald, P. B., Voineskos, A. N., Schutter, D. J., & Daskalakis, Z. J. (2013). Inhibitory deficits in the dorsolateral prefrontal cortex in psychopathic offenders. *Cortex, 49*(5), 1377–1385.

Hughes, G. M., Teeling, E. C., & Higgins, D. G. (2014). Loss of olfactory receptor function in hominin evolution. *PloS One, 9*(1), e84714.

Hurford, J. R. (2014). *Origins of language: A slim guide.* New York, NY: Oxford University Press.

Iacoboni, M., & Dapretto, M. (2006). The mirror neuron system and the consequences of its dysfunction. *Nature Reviews Neuroscience, 7*(12), 942–951.

Isaacson, R. L. (1972). Hippocampal destruction in man and other animals. *Neuropsychologia, 10,* 47–64.

Ishikawa, T., Tomatsu, S., Izawa, J., & Kakei, S. (2016). The cerebro-cerebellum: Could it be loci of forward models? *Neuroscience Research, 104,* 72–79.

Ito, M. (1970). Neurophysiological aspects of the cerebellar motor control system. *International Journal of Neurology, 7*(2), 162–176.

Ito, M. (1984). *The cerebellum and neural control.* New York, NY: Raven Press.

Ito, M. (1993). Movement and thought: Identical control mechanisms by the cerebellum. *Trends in Neuroscience, 16*(11), 448–450.

Ito, M. (2002). Historical review of the significance of the cerebellum and the role of Purkinje cells in motor learning. *Annals of the New York Academy of Sciences, 978*(1), 273–288.

Ito, M. (2008). Control of mental activities by internal models in the cerebellum. *Nature Reviews Neuroscience, 9*(4), 304–313.

Ito, A., Abe, N., Fujii, T., Hayashi, A., Ueno, A., Mugikura, S., . . . Mori, E. (2012). The contribution of the dorsolateral prefrontal cortex to the preparation for deception and truth-telling. *Brain Research, 1464,* 43–52.

Jackendoff, R. (1999). Possible stages in the evolution of the language capacity. *Trends in Cognitive Sciences, 3*(7), 272–279.

James, W. (1890). *The principles of psychology.* New York, NY: Henry Holt and Company.

Jenkins, J. G., & Dallenbach, K. M. (1924). Obliviscence during sleep and waking. *American Journal of Psychology, 35*(4), 605–612.

Jensen, P. S., Mrazek, D., Knapp, P. K., Steinberg, L., Pfeffer, C., Schowalter, J., & Shapiro, T. (1997). Evolution and revolution in child psychiatry: ADHD as a disorder of adaptation. *Journal of the American Academy of Child & Adolescent Psychiatry, 36*(12), 1672–1681.

Johanson, D. E., Johanson, D. C., Edgar, B., & Blake, E. (1996). *From Lucy to language.* New York, NY: Simon and Schuster.

Jouvet, M. (1972). The role of monoamines and acetylcholine-containing neurons in the regulation of the sleep-waking cycle. *Neurophysiology and Neurochemistry of Sleep and Wakefulness, 64,* 166–307.

Jouvet, M. (1980). Paradoxical sleep and the nature–nurture controversy. *Progress in Brain Research, 53,* 331–346.

Juric, I., Aeschbacher, S., & Coop, G. (2016). The strength of selection against Neanderthal introgression. *PLoS Genetics, 12*(11), e1006340.Kane, M. J., Hambrick, D. Z., & Conway, A. R. (2005). Working memory capacity and fluid intelligence are strongly related constructs: Comment on Ackerman, Beier, and Boyle (2005). *Psychological Bulletin, 131*(1), 66–71.

Kauffman, S. (2000). *Investigations.* New York, NY: Oxford University Press.

Kavanau, J. L. (2002). REM and NREM sleep as natural accompaniments of the evolution of warm-bloodedness. *Neuroscience & Biobehavioral Reviews, 26*(8), 889–906.

Keysers, C., & Gazzola, V. (2010). Social neuroscience: mirror neurons recorded in humans. *Current Biology, 20*(8), R353–R354.

Kheradmand, A., Lasker, A., & Zee, D. S. (2013). Transcranial magnetic stimulation (TMS) of the supramarginal gyrus: A window to perception of upright. *Cerebral Cortex, 25*(3), 765–771.

Kochiyama, T., Ogihara, N., Tanabe, H. C., Kondo, O., Amano, H., Hasegawa, K., . . . Stringer, C. (2018). Reconstructing the Neanderthal brain using computational anatomy. *Scientific Reports, 8*(1), 6296.

Köhler, W. (1929). *Gestalt psychology.* New York, NY: Liveright.

Kohn, M., & Mithen, S. (1999). Handaxes: Products of sexual selection? *Antiquity, 73*(281), 518–526.

Korsakoff, S. S. (1887). Disturbance of psychic function in alcoholic paralysis and its relation to the disturbance of the psychic sphere in multiple neuritis of non-alcoholic origin. *Vestnik Psichiatrii, 4*(2), 1–102.

Kozbelt, A. (2019). Evolutionary explanations for humor and creativity. In S. R. Luria, J. Baer, & J. C. Kaufman (Eds.), *Creativity and humor* (pp. 205–230). Cambridge, MA: Academic Press.

Koziol, L. F., Budding, D., Andreasen, N., D'arrigo, S., Bulgheroni, S., Imamizu, H., & Pezzulo, G. (2014). Consensus paper: The cerebellum's role in movement and cognition. *Cerebellum, 13*(1), 151–177.

Krippner, S., & Hughes, W. (1970). Genius at work. *Psychology Today, 4*(1), 40–43.

Kubo, D., Tanabe, H. C., Kondo, O., Ogihara, N., Yogi, A., Murayama, S., & Ishida, H. (2014). Cerebellar size estimation from endocranial measurements: An evaluation based on MRI data. In T. Akazawa, N. Ogihara, H. C. Tanabe, & H. Terashima (Eds.), *Dynamics of learning in Neanderthals and modern humans* (Vol. 2, pp. 209–215). Tokyo, Japan: Springer.

Kutter, E. F., Bostroem, J., Elger, C. E., Mormann, F., & Nieder, A. (2018). Single neurons in the human brain encode numbers. *Neuron, 100*(3), 753–761.

Laing, R. D. (1967). *The politics of experience.* New York, NY: Pantheon Books.

Leakey, L. S., & Leakey, M. D. (1964). Recent discoveries of fossil hominids in Tanganyika: At Olduvai and near Lake Natron. *Nature, 202*(4927), 5–7.

LeDoux, J. (1996). Emotional networks and motor control: A fearful view. *Progress in Brain Research, 107*, 437–446.

Leiner, H. C., Leiner, A. L., & Dow, R. S. (1986). Does the cerebellum contribute to mental skills? *Behavioral Neuroscience, 100*(4), 443–454.

Leiner, H. C., Leiner, A. L., & Dow, R. S. (1989). Reappraising the cerebellum: What does the hindbrain contribute to the forebrain? *Behavioral Neuroscience, 103*(5), 998–1008.

Leiner, H. C., Leiner, A. L., & Dow, R. S. (1991). The human cerebro-cerebellar system: Its computing, cognitive, and language skills. *Behavioural Brain Research, 44*(2), 113–128.

Levulis, L. I., & Coolidge, F. L. (2016, October). *Of moles and men: On the evolutionary implications of expanded olfactory bulbs in* Homo sapiens. A poster presented at the meeting of the European Society for the study of Human Evolution, Madrid, Spain.

Lewis, D. M., Al-Shawaf, L., Conroy-Beam, D., Asao, K., & Buss, D. M. (2017). Evolutionary psychology: A how-to guide. *American Psychologist, 72*(4), 353–373.

Lewis-Williams, J. D. (2002). *A cosmos in stone: Interpreting religion and society through rock art.* Walnut Creek, CA: AltaMira Press.

Lezak, M. D. (1982). The problem of assessing executive functions. *International Journal of Psychology, 17*(1-4), 281–297.

Lieberman, D. E., Raichlen, D. A., Pontzer, H., Bramble, D. M., & Cutright-Smith, E. (2006). The human gluteus maximus and its role in running. *Journal of Experimental Biology, 209*(11), 2143–2155.

Llinas, R. (1975). The cortex of the cerebellum. *Scientific American, 232*(1), 56–71.

Lou, H. C., Luber, B., Crupain, M., Keenan, J. P., Nowak, M., Kjaer, T. W., . . . Lisanby, S. H. (2004). Parietal cortex and representation of the mental self. *Proceedings of the National Academy of Sciences of the United States of America, 101*(17), 6827–6832.

Luria, A. R. (1966). *Higher cortical functions in man* (B. Haigh, Trans.). New York, NY: Basic Books. (Originally published 1962 by Moscow University Press)

MacCabe, J. H., Sariaslan, A., Almqvist, C., Lichtenstein, P., Larsson, H., & Kyaga, S. (2018). Artistic creativity and risk for schizophrenia, bipolar disorder and unipolar depression: A Swedish population-based case–control study and sib-pair analysis. *British Journal of Psychiatry, 212*(6), 370–376.

Mantini, D., Gerits, A., Nelissen, K., Durand, J. B., Joly, O., Simone, L., . . . Vanduffel, W. (2011). Default mode of brain function in monkeys. *Journal of Neuroscience, 31*(36), 12954–12962.

Margulies, D. S., Vincent, J. L., Kelly, C., Lohmann, G., Uddin, L. Q., Biswal, B. B., . . . Petrides, M. (2009). Precuneus shares intrinsic functional architecture in humans and monkeys. *Proceedings of the National Academy of Sciences of the United States of America, 106*(47), 20069–20074.

Marshall, L., & Born, J. (2007). The contribution of sleep to hippocampus-dependent memory consolidation. *Trends in Cognitive Sciences, 11*(10), 442–450.

Martin-Soelch, C., Linthicum, J., & Ernst, M. (2007). Appetitive conditioning: Neural bases and implications for psychopathology. *Neuroscience & Biobehavioral Reviews, 31*(3), 426–440.

McCandless, B. D., Cohen, L., & Dehaene, S. (2003). The visual word form area: Expertise for reading in the fusiform gyrus. *Trends in Cognitive Sciences, 7*(7), 293–299.

McCoy, R. C., Wakefield, J., & Akey, J. M. (2017). Impacts of Neanderthal-introgressed sequences on the landscape of human gene expression. *Cell, 168*(5), 916–927.

McGrew, W. C., & Marchant, L. F. (1997). On the other hand: Current issues in and meta-analysis of the behavioral laterality of hand function in nonhuman primates. *American Journal of Physical Anthropology, 104*(s25), 201–232.

McPherron, S. P., Alemseged, Z., Marean, C. W., Wynn, J. G., Reed, D., Geraads, D., . . . Béarat, H. A. (2010). Evidence for stone-tool-assisted consumption of animal tissues before 3.39 million years ago at Dikika, Ethiopia. *Nature, 466*(7308), 857–860.

Meehl, P. E. (1962). Schizotaxia, schizotypy, schizophrenia. *American Psychologist, 17*(12), 827–838.

Mendez, F. L., Poznik, G. D., Castellano, S., & Bustamante, C. D. (2016). The divergence of Neandertal and modern human Y chromosomes. *American Journal of Human Genetics, 98*, 728–734.

Meshberger, F. L. (1990). An interpretation of Michelangelo's *Creation of Adam* based on neuroanatomy. *JAMA: Journal of the American Medical Association, 264*(14), 1837–1841.

Mikhail, C., Vaucher, A., Jimenez, S., & Tafti, M. (2017). ERK signaling pathway regulates sleep duration through activity-induced gene expression during wakefulness. *Science Signaling, 10*(463), eaai9219.

Miller, G., Tybur, J. M., & Jordan, B. D. (2007). Ovulatory cycle effects on tip earnings by lap dancers: Economic evidence for human estrus? *Evolution and Human Behavior, 28*(6), 375–381.

Millon, T., Millon, C. M., Meagher, S. E., Grossman, S. D., & Ramnath, R. (2012). *Personality disorders in modern life.* New York, NY: John Wiley & Sons.

Mithen, S. (1996). *The prehistory of mind.* London, England: Thames and Hudson.

Moberget, T., & Ivry, R. B. (2016). Cerebellar contributions to motor control and language comprehension: Searching for common computational principles. *Annals of the New York Academy of Sciences, 1369*(1), 154–171.

Morrison, A. (1983). A window on the sleeping brain. *Scientific American, 248,* 94–102.

Moser, D., Anderer, P., Gruber, G., Parapatics, S., Loretz, E., Boeck, M., . . . Saletu, B. (2009). Sleep classification according to AASM and Rechtschaffen & Kales: Effects on sleep scoring parameters. *Sleep, 32*(2), 139–149.

Moser, M. B., & Moser, E. I. (1998). Functional differentiation in the hippocampus. *Hippocampus, 8*(6), 608–619.

Mujica-Parodi, L. R., Strey, H. H., Frederick, B., Savoy, R., Cox, D., Botanov, Y., . . . Weber, J. (2009). Chemosensory cues to conspecific emotional stress activate amygdala in humans. *PLoS One, 4*(7), e6415.

Nagel, T. (1974). What is it like to be a bat? *The Philosophical Review, 83*(4), 435–450.

Nesse, R. M. (1984). An evolutionary perspective on psychiatry. *Comprehensive Psychiatry, 25*(6), 575–580.

Neubauer, S., Hublin, J.-J., & Gunz, P. (2018). The evolution of modern human brain shape. *Science Advances, 4,* eaao5961.

Nichols, C. (2009). Is there an evolutionary advantage of schizophrenia? *Personality and Individual Differences, 46*(8), 832–838.

O'Keefe, J., & Burgess, N. (2005). Dual phase and rate coding in hippocampal place cells: Theoretical significance and relationship to entorhinal grid cells. *Hippocampus, 15*(7), 853–866.

Orban, G. A., & Caruana, F. (2014). The neural basis of human tool use. *Frontiers in Psychology, 5,* Article 310.

Oring, E. (1992). *Jokes and their relations.* Lexington, KY: University Press of Kentucky.

Osaka, N., Logie, R. H., & D'Esposito, M. (Eds.). (2007). *The cognitive neuroscience of working memory.* New York, NY: Oxford University Press.

Overmann, K. A., & Coolidge, F. L. (2013). Human species and mating systems: Neandertal–*Homo sapiens* reproductive isolation and the archaeological and fossil records. *Journal of Anthropological Sciences, 91,* 91–110.

Passingham, R. E., & Wise, S. P. (2012). *The neurobiology of the prefrontal cortex: Anatomy, evolution, and the origin of insight.* Oxford, UK: Oxford University Press.

Perls, F. S. (1969). *Gestalt therapy verbatim.* Moab, UT: Real People Press.

Pinker, S., & Jackendoff, R. (2005). The faculty of language: What's special about it? *Cognition, 95*(2), 201–236.

Polanczyk, G., de Lima, M. S., Horta, B. L., Biederman, J., & Rohde, L. A. (2007). The worldwide prevalence of ADHD: A systematic review and meta-regression analysis. *American Journal of Psychiatry, 164*(6), 942–948.

Poulos, A. M., & Thompson, R. F. (2015). Localization and characterization of an essential associative memory trace in the mammalian brain. *Brain Research, 1621*, 252–259.

Power, R. A., Steinberg, S., Bjornsdottir, G., Rietveld, C. A., Abdellaoui, A., Nivard, M. M., . . . Cesarini, D. (2015). Polygenic risk scores for schizophrenia and bipolar disorder predict creativity. *Nature Neuroscience, 18*(7), 953–955.

Price, T., Wadewitz, P., Cheney, D., Seyfarth, R., Hammerschmidt, K., & Fischer, J. (2015). Vervets revisited: A quantitative analysis of alarm call structure and context specificity. *Scientific Reports, 5*, 13220.

Pulvermüller, F. (2018). Neural reuse of action perception circuits for language, concepts and communication. *Progress in Neurobiology, 160*, 1–44.

Raizen, D. M., Zimmerman, J. E., Maycock, M. H., Ta, U. D., You, Y. J., Sundaram, M. V., & Pack, A. I. (2008). Lethargus is a *Caenorhabditis elegans* sleep-like state. *Nature, 451*(7178), 569–572.

Ramachandran, V. S., & Hubbard, E. M. (2001). Synaesthesia—A window into perception, thought and language. *Journal of Consciousness Studies, 8*(12), 3–34.

Rampin, C., Cespuglio, R., Chastrette, N., & Jouvet, M. (1991). Immobilisation stress induces a paradoxical sleep rebound in rat. *Neuroscience Letters, 126*(2), 113–118.

Rangarajan, V., Hermes, D., Foster, B. L., Weiner, K. S., Jacques, C., Grill-Spector, K., & Parvizi, J. (2014). Electrical stimulation of the left and right human fusiform gyrus causes different effects in conscious face perception. *Journal of Neuroscience, 34*(38), 12828–12836.

Rechtschaffen, A. (1971). The control of sleep. In W. A. Hunt (Ed.), *Human behavior and its control.* Oxford, UK: Schenkman.

Rechtschaffen, A., & Kales, A. (1968). *A manual of standardized terminology, techniques, and scoring systems for sleep stages of human subjects.* Washington, DC: US Government Printing Office.

Revonsuo, A. (2000). The reinterpretation of dreams: An evolutionary hypothesis of the function of dreaming. *Behavioral and Brain Sciences, 23*(6), 877–901.

Revonsuo, A., Tuominen, J., & Valli, K. (2015). The avatars in the machine: Dreaming as a simulation of social reality. In *Open MIND.* Frankfurt am Main, Germany: MIND Group.

Ribeiro, S., Goyal, V., Mello, C. V., & Pavlides, C. (1999). Brain gene expression during REM sleep depends on prior waking experience. *Learning & Memory, 6*(5), 500–508.

Rilling, J. K. (2014). Comparative primate neuroimaging: Insights into human brain evolution. *Trends in Cognitive Sciences, 18*(1), 46–55.

Rizzolatti, G., & Craighero, L. (2004). The mirror-neuron system. *Annual Review of Neuroscience, 27*, 169–192.

Robins, L. N. (1966). *Deviant children grown up: A sociological and psychiatric study of sociopathic personality.* Baltimore, MD: Williams & Wilkins.

Rossano, M. J. (2010). *Cautiously digging up the mind: Cognitive archeology and human evolution.* Cambridge, UK: Cambridge University Press.

Rougier, H., Crevecoeur, I., Beauval, C., Posth, C., Flas, D., Wißing, C., . . . van der Plicht, J. (2016). Neandertal cannibalism and Neandertal bones used as tools in Northern Europe. *Scientific Reports, 6*, 29005.

Ruff, C. B., & Burgess, M. L. (2015). How much more would KNM-WT 15000 have grown? *Journal of Human Evolution, 80*, 74–82.

Sagan, C. (1977). *The dragons of Eden: Speculations on the evolution of human intelligence.* New York, NY: Random House.

Sagan, C. (1996). *The demon-haunted world: Science as a candle in the dark.* New York, NY: Random House.

Saggar, M., Quintin, E. M., Kienitz, E., Bott, N. T., Sun, Z., Hong, W. C., . . . Hawthorne, G. (2015). Pictionary-based fMRI paradigm to study the neural correlates of spontaneous improvisation and figural creativity. *Scientific Reports, 5*, 10894.

Sawyer, S., Renaud, G., Viola, B., Hublin, J. J., Gansauge, M. T., Shunkov, M. V., . . . Pääbo, S. (2015). Nuclear and mitochondrial DNA sequences from two Denisovan individuals. *Proceedings of the National Academy of Sciences of the United States of America, 112*(51), 15696–15700.

Schaaffhausen, M. (1865). Sur l'opuscule de M. Fuhlrott sur l'homme fossile de Néanderthal, *Bulletins De La Société Danthropologie De Paris, 6*(1), 688–690.

Schacter, D. L. (2012). Adaptive constructive processes and the future of memory. *American Psychologist, 67*(8), 603–613.

Schacter, D. L., & Addis, D. R. (2007). The cognitive neuroscience of constructive memory: Remembering the past and imagining the future. *Philosophical Transactions of the Royal Society B: Biological Sciences, 362*(1481), 773–786.

Schacter, D. L., Guerin, S. A., & St. Jacques, P. L. (2011). Memory distortion: An adaptive perspective. *Trends in Cognitive Sciences, 15*(10), 467–474.

Schiller, F. (1992). *Paul Broca: Founder of French anthropology, explorer of the brain.* New York, NY: Oxford University Press.

Schizophrenia Working Group of the Psychiatric Genomics Consortium. (2014). Biological insights from 108 schizophrenia-associated genetic loci. *Nature, 511*(7510), 421–427.

Schlenker, P., Chemla, E., Arnold, K., & Zuberbühler, K. (2016). Pyow-hack revisited: Two analyses of putty-nosed monkey alarm calls. *Lingua, 171*, 1–23.

Schlerf, J. E., Verstynen, T. D., Ivry, R. B., & Spencer, R. M. (2010). Evidence of a novel somatotopic map in the human neocerebellum during complex actions. *Journal of Neurophysiology, 103*(6), 3330–3336.

Schmahmann, J. D. (1991). An emerging concept: The cerebellar contribution to higher function. *Archives of Neurology, 48*(11), 1178–1187.

Schmahmann, J. D. (2004). Disorders of the cerebellum: Ataxia, dysmetria of thought, and the cerebellar cognitive affective syndrome. *Journal of Neuropsychiatry and Clinical Neurosciences, 16*(3), 367–378.

Schmajuk, N. A., Lam, Y. W., & Gray, J. A. (1996). Latent inhibition: A neural network approach. *Journal of Experimental Psychology: Animal Behavior Processes, 22*(3), 321.

Scott-Phillips, T. C., & Blythe, R. A. (2013). Why is combinatorial communication rare in the natural world, and why is language an exception to this trend? *Journal of the Royal Society Interface, 10*(88), 1–7.

Segal, D. L., Coolidge, F. L., & Rosowsky, E. (2006). *Personality disorders and older adults: Diagnosis, assessment, and treatment.* New York, NY: Wiley.

Seghier, M. L. (2013). The angular gyrus: Multiple functions and multiple subdivisions. *The Neuroscientist, 19*(1), 43–61.

Shah, P., & Miyake, A. (Eds.). (2005). *The Cambridge handbook of visuospatial thinking.* Cambridge, MA: Cambridge University Press.

Shepard, R. (1997). The genetic basis of human scientific knowledge. In G. Bock & G. Cardew (Eds.), *Characterizing human psychological adaptations* (pp. 4–13). Chichester, England: Wiley and Sons.

Shermer, M. (2011). *The believing brain: From ghosts and gods to politics and conspiracies— How we construct beliefs and reinforce them as truths.* New York, NY: Macmillan.

Shipton, C. (2010). Imitation and shared intentionality in the Acheulean. *Cambridge Archaeological Journal, 20*(02), 197–210.

Siegel, J. M. (2009). Sleep viewed as a state of adaptive inactivity. *Nature Reviews Neuroscience, 10*(10), 747–753.

Silani, G., Lamm, C., Ruff, C. C., & Singer, T. (2013). Right supramarginal gyrus is crucial to overcome emotional egocentricity bias in social judgments. *Journal of Neuroscience, 33*(39), 15466–15476.

Silber, M. H., Ancoli-Israel, S., Bonnet, M. H., Chokroverty, S., Grigg-Damberger, M. M., Hirshkowitz, M., . . . Iber, C. (2007). The visual scoring of sleep in adults. *Journal of Clinical Sleep Medicine, 3*(2), 121–131.

Silverman, J. (1967). Shamans and acute schizophrenia. *American Anthropologist, 69*(1), 21–31.

Singh, M. (2018). The cultural evolution of shamanism. *Behavioral and Brain Sciences, 41*, 1–62.

Skinner, B. F. (1948). *Walden two.* New York, NY: Macmillan.

Skinner, B. F. (1971). *Beyond freedom and dignity.* New York, NY: Random House.

Slon, V., Mafessoni, F., Vernot, B., de Filippo, C., Grote, S., Viola, B., . . . Douka, K. (2018). The genome of the offspring of a Neanderthal mother and a Denisovan father. *Nature, 561*(7721), 113–116.

Solstad, T., Boccara, C. N., Kropff, E., Moser, M. B., & Moser, E. I. (2008). Representation of geometric borders in the entorhinal cortex. *Science, 322*(5909), 1865–1868.

Spoor, F., Gunz, P., Neubauer, S., Stelzer, S., Scott, N., Kwekason, A., & Dean, M. C. (2015). Reconstructed *Homo habilis* type OH 7 suggests deep-rooted species diversity in early *Homo. Nature, 519*(7541), 83–86.

Spreng, R. N., Stevens, W. D., Chamberlain, J. P., Gilmore, A. W., & Schacter, D. L. (2010). Default network activity, coupled with the frontoparietal control network, supports goal-directed cognition. *NeuroImage, 53*(1), 303–317.

Srinivasan, S., Bettella, F., Mattingsdal, M., Wang, Y., Witoelar, A., Schork, A. J., . . . Collier, D. A. (2016). Genetic markers of human evolution are enriched in schizophrenia. *Biological Psychiatry, 80*(4), 284–292.

Stalnaker, T. A., Cooch, N. K., & Schoenbaum, G. (2015). What the orbitofrontal cortex does not do. *Nature Neuroscience, 18*(5), 620–627.

Stickgold, R., Scott, L., Rittenhouse, C., & Hobson, J. A. (1999). Sleep-induced changes in associative memory. *Journal of Cognitive Neuroscience, 11*(2), 182–193.

Stickgold, R., & Walker, M. P. (2013). Sleep-dependent memory triage: Evolving generalization through selective processing. *Nature Neuroscience, 16*, 139–145.

Stout, D., & Chaminade, T. (2012). Stone tools, language and the brain in human evolution. *Philosophical Transactions of the Royal Society B, 367*(1585), 75–87.

Stout, D., Hecht, E., Khreisheh, N., Bradley, B., & Chaminade, T. (2015). Cognitive demands of Lower Paleolithic toolmaking. *PloS One, 10*(4), e0121804.

Sukhoverkhov, A. V., & Fowler, C. A. (2015). Why language evolution needs memory: Systems and ecological approaches. *Biosemiotics, 4*(8), 47–65.

Tate, M. C., Herbet, G., Moritz-Gasser, S., Tate, J. E., & Duffau, H. (2014). Probabilistic map of critical functional regions of the human cerebral cortex: Broca's area revisited. *Brain*, *137*(10), 2773–2782.

Thorndike, E. (1905). *The elements of psychology.* New York, NY: A. G. Seiler.

Thornhill, R., & Gangestad, S. W. (2008). *The evolutionary biology of human female sexuality.* Oxford, UK: Oxford University Press.

Tononi, G., & Cirelli, C. (2014). Sleep and the price of plasticity: From synaptic and cellular homeostasis to memory consolidation and integration. *Neuron*, *81*(1), 12–34.

Torgersen, S. (2000). Genetics of patients with borderline personality disorder. *Psychiatric Clinics of North America*, *23*(1), 1–9.

Torgersen, S. (2009). The nature (and nurture) of personality disorders. *Scandinavian Journal of Psychology*, *50*(6), 624–632.

Torgersen, S., Lygren, S., Oien, P. A., Skre, I., Onstad, S., Edvardsen, J., Tambs, K., & Kringlen, E. (2000). A twin study of personality disorders. *Comprehensive Psychiatry*, *41*, 416–425.

Toth, N. (1985). Archaeological evidence for preferential right-handedness in the Lower and Middle Pleistocene, and its possible implications. *Journal of Human Evolution*, *14*(6), 607–614.

Tottenham, N., & Sheridan, M. A. (2010). A review of adversity, the amygdala and the hippocampus: A consideration of developmental timing. *Frontiers in Human Neuroscience*, *3*, 68.

Trivers, R. (1972). *Parental investment and sexual selection* (Vol. *136*). Cambridge, MA: Biological Laboratories, Harvard University.

Tucker, M. A., Nguyen, N., & Stickgold, R. (2016). Experience playing a musical instrument and overnight sleep enhance performance on a sequential typing task. *PloS One*, *11*(7), e0159608.

Tulving, E. (1972). Episodic and semantic memory. In E. Tulving & W. Donaldson (Eds.), *Organization of memory* (pp. 382–423). Oxford, UK: Academic Press.

Tulving, E. (1995). Memory: Introduction. In M. Gazzaniga (Ed.), *The cognitive sciences* (pp. 751–753). Cambridge, MA: MIT Press.

Tulving, E. (2002). Episodic memory: From mind to brain. *Annual Review of Psychology*, *53*(1), 1–25.

Turkheimer, E. (2000). Three laws of behavior genetics and what they mean. *Current Directions in Psychological Science*, *9*(5), 160–164.

Uomini, N. T. (2009). The prehistory of handedness: Archaeological data and comparative ethology. *Journal of Human Evolution*, *57*(4), 411–419.

Van de Castle, R. (1994). *Our dreaming mind: A sweeping exploration of the role that dreams have played in politics, art, religion, and psychology, from ancient civilizations to the present day.* New York, NY: Ballantine Books.

Vandervert, L. (2003). How working memory and cognitive modeling functions of the cerebellum contribute to discoveries in mathematics. *New Ideas in Psychology*, *21*, 159–175.

Vandervert, L. (2015). How music training enhances working memory: A cerebrocerebellar blending mechanism that can lead equally to scientific discovery and therapeutic efficacy in neurological disorders. *Cerebellum & Ataxias*, *2*, 11.

Vandervert, L. (2016). The prominent role of the cerebellum in the learning, origin and advancement of culture. *Cerebellum & Ataxias*, *3*, 10.

Vandervert, L. R., Schimpf, P. H., & Liu, H. (2007). How working memory and the cerebellum collaborate to produce creativity and innovation. *Creativity Research Journal*, *19*(1), 1–18.

Van Horn, J. D., Irimia, A., Torgerson, C. M., Chambers, M. C., Kikinis, R., & Toga, A. W. (2012). Mapping connectivity damage in the case of Phineas Gage. *PloS One, 7*(5), e37454.

Vann, S. D., Aggleton, J. P., & Maguire, E. A. (2009). What does the retrosplenial cortex do? *Nature Reviews Neuroscience, 10*(11), 792–802.

Van Overwalle, F., Baetens, K., Mariën, P., & Vandekerckhove, M. (2014). Social cognition and the cerebellum: A meta-analysis of over 350 fMRI studies. *NeuroImage, 86*, 554–572.

Villa, P., & Roebroeks W. (2014). Neandertal demise: An archaeological analysis of the modern human superiority complex. *PLoS One, 9*, 1–10.

Wagner, U., Gais, S., Haider, H., Verleger, R., & Born, J. (2004). Sleep inspires insight. *Nature, 427*(6972), 352–355.

Walker, A., & Leakey, R. E. (Eds.). (1993). *The Nariokotome* Homo erectus *skeleton*. Cambridge, MA: Harvard University Press.

Walker, M. P. (2005). A refined model of sleep and the time course of memory formation. *Behavioral and Brain Sciences, 28*(1), 51–64.

Weaver, A. H. (2005). Reciprocal evolution of the cerebellum and neocortex in fossil humans. *Proceedings of the National Academy of Sciences of the United States of America, 102*(10), 3576–3580.

Weaver, A. H. (2010). Cerebellum and brain evolution in Holocene humans. In D. Broadfield, M. Yuan, K. Schick, & N. Toth (Eds.), *The human brain evolving: Paleoneurological studies in honor of Ralph. L. Holloway* (pp. 97–106). Bloomington, IN: Stone Age Institute.

Weiner, K. S., & Zilles, K. (2016). The anatomical and functional specialization of the fusiform gyrus. *Neuropsychologia, 83*, 48–62.

Wells, J. C., & Stock, J. T. (2007). The biology of the colonizing ape. *American Journal of Physical Anthropology, 134*(S45), 191–222.

Wernicke, C. (1874). *The aphasic symptom complex: A psychological study on an anatomical basis*. Breslau, Germany: Max Cohn & Weigert.

White, T. D., Asfaw, B., DeGusta, D., Gilbert, H., Richards, G. D., Suwa, G., & Howell, F. C. (2003). Pleistocene *Homo sapiens* from Middle Awash, Ethiopia. *Nature, 423*(6941), 742–747.

Wilson, E. O. (2012). *On human nature*. Cambriuge, MA: Harvard University Press.

Wing, Y. K., Lee, S. T., & Chen, C. N. (1994). Sleep paralysis in Chinese: Ghost oppression phenomenon in Hong Kong. *Sleep: Journal of Sleep Research & Sleep Medicine, 17*, 609–613.

Winson, J. (1990). The meaning of dreams. *Scientific American, 263*, 89–96.

Wittgenstein, L. (1953). *Philosophical investigations*. (G. E. M. Anscombe, Trans.). Oxford, UK: Blackwell.

Wynn, T. (2002). Archaeology and cognitive evolution. *Behavioral and Brain Sciences, 25*(3), 389–402.

Wynn, T., & Coolidge, F. L. (2007a). A Stone-Age meeting of minds. *American Scientist, 96*, 44–51.

Wynn, T., & Coolidge, F. L. (2007b). Did a small but significant enhancement in working memory capacity power the evolution of modern thinking? In P. Mellars, K. Boyle,

O. Bar-Yosef, & S. Stringer (Eds.), *Rethinking the human evolution: New behavioural and biological perspectives on the origin and dispersal of modern humans* (pp. 79–90). Cambridge, UK: Cambridge University McDonald Institute Monographs.

Wynn, T., & Coolidge, F. L. (2010). Beyond symbolism and language. In T. Wynn, & F. L. Coolidge (Eds.), *Working memory: Beyond language and symbolism* (pp. S5–S16). Chicago, IL: University of Chicago Press.

Wynn, T., & Coolidge, F. L. (2012). *How to think like a Neandertal.* Oxford, UK: Oxford University Press.

Wynn, T., & Coolidge, F. L. (2016). Archeological insights into hominin cognitive evolution. *Evolutionary Anthropology: Issues, News, and Reviews, 25*(4), 200–213.

Wynn, T., & Gowlett, J. (2018). The handaxe reconsidered. *Evolutionary Anthropology: Issues, News, and Reviews, 27*(1), 21–29.

Wynn, T., Overmann, K. A., & Coolidge, F. L. (2016). The false dichotomy: A refutation of the Neandertal indistinguishability claim, 1–22.

Xie, L., Kang, H., Xu, Q., Chen, M.J., Liao, Y., Thiyagarajan, M., . . . Nedergaard M. (2013). Sleep drives metabolite clearance from the adult brain. *Science, 342*(6156), 373–377.

Yuen, R. K., Merico, D., Bookman, M., Howe, J. L., Thiruvahindrapuram, B., Patel, R. V., . . . Pellecchia, G. (2017). Whole genome sequencing resource identifies 18 new candidate genes for autism spectrum disorder. *Natural Neuroscience, 20*(4), 602–611.

Zaehle, T., Jordan, K., Wüstenberg, T., Baudewig, J., Dechent, P., & Mast, F. W. (2007). The neural basis of the egocentric and allocentric spatial frame of reference. *Brain Research, 1137*, 92–103.

Zhang, R., Lahens, N. F., Ballance, H. I., Hughes, M. E., & Hogenesch, J. B. (2014). A circadian gene expression atlas in mammals: Implications for biology and medicine. *Proceedings of the National Academy of Sciences of the United States of America, 111*(45), 16219–16224.

Zilhão J. (2014). Neandertal–modern human contact in Western Eurasia: Issues of dating, taxonomy, and cultural associations. In T. Akazawa, Y. Nishiak, & K. Aoki (Eds.), *Dynamics of learning in Neanderthals and modern humans: Cultural perspectives* (Vol. 1, pp. 21–57). Tokyo: Springer.

Zorzi, M., Di Bono, M. G., & Fias, W. (2011). Distinct representations of numerical and non-numerical order in the human intraparietal sulcus revealed by multivariate pattern recognition. *NeuroImage, 56*(2), 674–680.

Index

For the benefit of digital users, indexed terms that span two pages (e.g., 52–53) may, on occasion, appear on only one of those pages.

Figures are indicated by *f* following the page number

Aboitiz, F., 11, 44–45, 120–21
abstraction
  cerebellum in, 165
  evolution of, 74, 105
acalculia, 72–73, 121
Acanthostega, 10–11
Acheulean handaxes. *See* handaxes
action perception circuits, 141–42
adaptations
  characterization, 1, 3, 25, 39–40
  of frontal lobes, 99–100
  of parietal lobes, 3, 128
adapted to flee famine
  hypothesis, 228–29
Addis, D. R., 126
ADHD, 1–3, 226
Aeschbacher, S., 21–22
Aggleton, J. P., 75–76
agnosia, 71, 114, 128
Agustí, J., 123–24
alexithymia, 80
Allen, T. A., 199
Al-Shawaf, L., 229, 230, 232
Alzheimer's disease, 170–71
amygdala
  in appetitive conditioning, 158
  characterization, 70f, 76, 82, 134–35,
    167f, 172, 176
  damage studies, 176–77
  hunger promotion/suppression, 174–75
  olfactory bulbs and, 169–70
Anderson, M., 5
angular gyrus, 5–6, 71f, 72–73, 114, 116f,
  121, 128, 134–35
anorexia nervosa, 228
anosmia, 92–93

anterior cingulate cortex
  in appetitive conditioning, 158
  characterization, 80
anterograde amnesia, 82
antisocial personality disorder, 212–14
anxiety/depression, 127–28
aphemia, 130–32
apophenia, 217–18
appetitive conditioning, 158
appropriate incongruity, 150
apraxia, 71
Ardipithecus ramidus, 12
Aristotle, 56–57, 152
Aron, A. R., 107–8
Aserinsky, E., 185–86
associative cognitive maps, 67–68
astrocytes characterization, 4–5
Auger, S. D., 125
Aurignacian culture, 130–32, 146–47
australopithecines, 18, 19–20, 51, 55–56,
  110, 195–96
Australopithecus afarensis (Lucy), 12, 13f,
  130, 138–39, 139f, 205–6, 227–28
autism spectrum disorders, 80, 100,
  140–41, 161–62
autoimmune diseases, 170–71
autonoesis, 74
autotopagnosia, 114
aversive conditioning, 158
avoidant personality disorders, 220
axons characterization, 4

Baddeley, A. D., 28, 41–44, 46, 49–51,
  105–6, 118–19
Baddeley's working memory model, 41f,
  41–46, 53, 105–6, 107, 118–19, 159–60